Windfall

The PRAIRIE WOMAN WHO
LOST HER WAY *and the*
GREAT-GRANDDAUGHTER
WHO FOUND HER

Erika Bolstad

This book is a memoir. It reflects the author's present recollections of experiences
over a period of time. Some names and characteristics have been changed, some
events have been compressed, and some dialogue has been re-created.

Published by Sourcebooks
P.O. Box 4410, Naperville, Illinois 60567–4410
(630) 961-3900
sourcebooks.com

Library of Congress Cataloging-in-Publication Data

Names: Bolstad, Erika, author.
Title: Windfall : the prairie woman who lost her way and the
    great-granddaughter who found her / Erika Bolstad.
Description: Naperville, Illinois : Sourcebooks, 2023. | Includes
    bibliographical references and index. | Summary: "Author Erika Bolstad
    was shocked to learn she had inherited mineral rights in North Dakota in
    the throes of an oil bonanza. Determined to unearth the story behind her
    unexpected inheritance, she followed the trail to her great-grandmother,
    Anna, who her family had painted to be a courageous homesteader who
    paved her way in the unforgiving American West. But, Bolstad discovers a
    darker truth about Anna than her family had ever shared. With
    journalistic rigor, she unearths a history of environmental exploitation
    and genocide as well as the modern-day consequences of the Great Plains
    Dream: we could be rich"-- Provided by publisher.
Identifiers: LCCN 2022030626 | (hardcover) | (epub) | (adobe pdf)
Subjects: LCSH: Bolstad, Erika. | Petroleum industry--North Dakota. |
    Mineral rights--North Dakota. | Family secrets.
Classification: LCC HD9567.N9 B65 2023 | DDC
    338.2/728209784--dc23/eng/20220715
LC record available at https://lccn.loc.gov/2022030626

Printed and bound in the United States of America.
VP 10 9 8 7 6 5 4 3 2 1

For Nan

"Of all the persistent qualities in American history, the values attached to property retain the most power."

—Patricia Nelson Limerick, *The Legacy of Conquest*

CONTENTS

Part III

Part I

1

FRACTURED

December 2009

North Dakota crude: *$63.96 per barrel*

Her name was Anna Josephine Sletvold. That's about all I knew when all of this began during the darkest days of the Great Recession.

Anna, my mother told me, was a plucky woman who, on her own, settled the untamed prairies of northwestern North Dakota. The family lore was made even more romantic for what it left out: Anna disappeared from her homestead in 1907, lost to time and the vast plains.

More than a century later, an oil company sent my mother a $2,400 check. The oil company was leasing mineral rights along the edges of the booming Bakken oil fields of North Dakota. From the oil company, my mother learned she was an heir to mineral rights below the surface of the land where Anna once had a homestead. The check arrived in a manila envelope a few days before Christmas in 2009, its auspicious timing further confirmation of a family theory, never articulated but well understood, that unexpected windfalls have a way of showing up when they are most welcome.

In 1951, when my mother was just six years old, her father had signed a lease with an oil company, during the first oil boom in North Dakota. It was on the windswept land where Anna, his mother, once staked her claim. The oil company never drilled on Anna's land, but it kept renewing the lease for more than a decade. All that lease money was enough to send my mother to college—she was the first person in her family to go.

My mother loved a windfall; how could she not? Her entire life, she'd heard the promises blowing across the Great Plains. She bought lottery tickets whenever the jackpots soared and scratch-offs on a whim. She stockpiled pocket change to play the slot machines at the Spirit Mountain Casino near the western Oregon town where she and my father raised us. "Do you know how long you can make ten dollars last on penny and five cent slots?" she once asked me. She took out loans, secure in the knowledge that there would, somehow, be money to pay them off when the time came. It had always been that way; it had always worked.

Someday, all this will be yours, my mother promised.[1]

We knew she was dying. A few months before the envelope arrived in her mail, an off-duty nurse found our mother passed out in a gym locker room from yet another heart attack. My sister, Stephanie, lived in Miami, and I lived in Washington, DC. By the time we both got to the hospital in Oregon, our mother was alert enough to ask us to bring in her jewelry. She told us she had picked out what she wanted us to have.

In my mother's office was a handwritten ledger on lined notebook paper detailing the $72,000 in medical debt she and my father owed to the hospital, to a cardiologist, and to the ambulance company. The oil company's $2,400 check barely made a dent.

Three months after she learned of her inheritance in North Dakota, my mother died.

The night of her death, I sat on the floor of my cold apartment in Washington, DC, the phone pressed to my ear as my father shared the news. My mother's caregivers, experienced with such transitions, could tell from the way she was breathing that the end was near. They urged my father to stay instead of going home for the night. My father wept as he described my mother's last few breaths. It was the first time I had ever heard him cry.

My grief came wrapped in guilt. I should have been there at the end. I knew it was the loneliest my father had ever been. But I couldn't afford to fly back to Oregon yet again. I had been there twice since my mother's most recent heart attack.

Shortly after the excitement of President Barack Obama's inauguration, the executives at the media company where I worked laid off hundreds of my colleagues across the country, and they cut my pay by more than $350 per month. I tried not to think too much about the collision course of my reality: I worked for a floundering newspaper chain crippled by the recession, and I rented a too-expensive Washington, DC, apartment. Until that point, my life had seemed to be on a steadily upward trajectory. One job always led to another better paying, more prestigious one.

But in 2009, I was never sure whether my rent check would clear, until one day it didn't. I barely had money to get my hair cut anymore let alone afford the honey-colored highlights that made me a little brighter, a little blonder.

The night of my mother's death, I pressed my back into the vertical lines of the radiator, trying to offset the hollowed-out feeling in my belly with warmth and some sort of sharp, physical sensation. The nubs of the carpet pressed into my bottom. I could see all the dog hair under my bed. I put on a favorite black turtleneck and kept it on for several days. It was a warm, cashmere cocoon that insulated me from the assault of a cold March and the ache of loss, deep in my gut.

The next day, I wrote my mother's obituary. Never had my words mattered more. As the writer in the family, it was the one thing I could do, and do well.

I had written plenty of obituaries over the years. Obituary writing was a rite of passage for young newspaper reporters working Saturday night shifts in otherwise quiet newsrooms. It was never easy to call the family after someone died, but it also wasn't as awful as you might imagine. I began the calls by bluffing my way into their lives a little, my voice warm without being phony.

"I'm so sorry for intruding at this time," I said.

Only once had someone said: "No you're not." But then they added, embarrassed, "I know you're just doing your job."

Mostly, people wanted to talk about the loved one they had just lost. Often they told extraordinary stories, even about the most ordinary lives. They only needed a little encouragement to share. In many places, the call from the newspaper was expected. Everyone understood that summing up a life in words was one of the few universal rituals left in modern American life.

I clung to that ritual. I sat on my tan velvet sofa with my laptop, wearing my favorite wide-legged sweatpants, my soft turtleneck, and the hammered-gold hoop earrings I pilfered from my mother's jewelry box on my last visit home. I wrote in a fury, unaware of time. An obituary, I understood, was an account of what was important to the people left behind. You could hit the highlights—my mother's long marriage to our father, how they met in college, how she was the granddaughter of a prairie homesteader—but it left out so much. I was telling the story of my mother as I knew it, as her daughter. It wasn't her story as she might have told it, but I'm certain it would have pleased her, because I wrote it for her. For the first time, I began to understand the limits of my profession. A story filtered through newspaper impartiality was not the same as an obituary told tempered by grief.

A few years later, when I was trying to become a mother, I read that fetal DNA lingers in a woman's bloodstream throughout her lifetime. Mothers always carry something of their children within their bodies. This, I finally understood, was why our grief for our mothers feels so raw, so physical. We hurt because those little pieces of our own DNA, the ones inside our mothers, died, too. Part of us is also gone forever, and our bodies know it.

The raw feeling subsided as I wrote the obituary. The hollow part remained.

The editor at one of the newspapers I wrote for, the *Idaho Statesman*, offered to print the obituary for free, even though my mother had never lived in Boise. It was an employee perk, and I took her up on it. The death notice ran as a news item in our hometown newspaper, the *Statesman Journal* in Salem, Oregon. But the full obituary cost $285.99.[2] I checked the balance on my American Express card. There was just enough room left on my credit limit. I would figure out later how to pay off the balance. Like mother, like daughter.

That night, my friends gathered in my small apartment. They brought pizza and burritos and beer. They wrapped me in the warm comfort of company, in the universal ritual of gathering together after a death. Their presence reminded me: I was not alone, I was loved, and I would continue to be loved.

———

Later that summer, my sister Stephanie and I traveled once again to Oregon, to spread my mother's ashes. We gathered at a county park, just up the road from the house where we once lived. The park was the scene of many happy summer days, in part for the shallow, kid-friendly rapids at a bend in the creek. We loved the natural rock chute, just slippery enough to slide down, even as it pilled and snagged the bottoms of our swimsuits and stained them green with moss. Steph's twin girls took a dizzy spin on the same lopsided merry-go-round we played on as kids, its revolutions as wobbly as ever. My five-year-old nephew, puzzled by the solemnity of the adults around him, squatted near the creek bank, peering into the shallows for fish.

I recorded on my iPhone the sound of the wind rustling the leaves of the cottonwoods that nestled along the banks of the creek. For a while after, the audio file showed up randomly whenever I played music in shuffle mode. The sound, always a puzzling interlude, returned me to the sun-dappled creek bottom the day we said goodbye.

Even now, when I hear wind in cottonwoods, I think of spreading my mother's ashes. Some sank to the bottom of the creek bed that day. Other granules floated, suspended in the current. As we released the ashes to the water, I thought of my mother's spirit floating down Willamina Creek to the Yamhill River, then to the Willamette River, and then into the wide Columbia. I liked to believe that some of those tiny motes reached the Pacific Ocean.

That evening, my sister and I sat at the smooth kitchen table our father had built from ash wood. We divided up our mother's jewelry. This was our physical inheritance, the objects our mother once wore that we would now wear, close to our skin. Steph, who wore a simple silver wedding band, chose our mother's diamond wedding ring. I picked out my mother's gold ginkgo-shaped earrings and an old-fashioned gold brooch inlaid with pearls. I had worn the brooch to my eighth-grade graduation, pinned to the lace, dropped-waist Jessica McClintock dress my mother bought me at the outlet store in San Francisco. The brooch had belonged to her grandmother, Anna, my mother once told me.

The oil company paperwork sat in a pile of papers on my father's kitchen counter. I read over the lease, trying to puzzle out how much money we might earn from royalties if the company ever exercised its option to drill for oil on land in a remote, northwest corner of North Dakota.

It seemed such an improbable windfall. We didn't even own the land, just the oil deep beneath the earth. Besides being born, what had we done to inherit mineral rights from a woman lost to the prairies and to history—until an oil company came calling one hundred years later? And who was Anna, really?

I held the brooch in my hand, my fingers rubbing the pearls. Anna had touched this object. She, too, had worn the brooch pinned next to her heart. She, too, must have once believed her land would bring her wealth.

A tiny whisper called to me at the kitchen table. It was a tendril of a story beckoning me to follow, the same whisper my mother had heard all her life: *We could be rich.*

My mother left me a mystery. It was my inheritance, my windfall. My story to tell.

2

PRAIRIE POTHOLES

August 2013

North Dakota crude: *$97.18 per barrel*

THE HORIZON WAS FLAT, BUT THE UNMARKED ROAD CURVY, CONFOUND-
ing everything I thought I knew about the North Dakota prairie. My
tires spun in the mud, working to gain purchase on ground softened
by the storm that had been letting up when I landed at the airport a
few hours earlier.

On the map, it looked like a road. My GPS insisted it was the
right direction. But as I drove toward my destination, the road seemed
more like a farm track, its ruts separated by a tall grass strip so high it
scraped the undercarriage of my rental car.

Panic made me focus. I didn't want to get mired in mud my first
day in North Dakota. I didn't want to ask for help, not that there was
anyone to ask, and even if there were, I was worried they might have
a shotgun. If this was a private road, I didn't want to be arrested for
trespassing. I didn't want to explain myself. I certainly didn't want to
look stupid for being there, wherever I actually was. This adventure
was too new to be judged by strangers.

I floored the accelerator, determined to conquer the muddy road. The tires spit gravel as the car skidded around a bend to safety on firmer ground. I pulled over to look at the map again, still uncertain of my exact whereabouts. Somewhere out there was a rectangular plot of land that my great-grandmother, Anna, had once claimed as her own.

By then, the rain had stopped and the skies were clearing. To each side of the car were fields of still-green wheat, sowed in neat, wavy rows that traced the contours of the terrain. Wild sunflowers lined the gravel road in chaotic yellow roadside glory. I rolled down my windows, only to be overwhelmed by the sound of peak summer cicada song. The air smelled fresh, like recent rain on top of mud, with a tinge of far-off manure. To the east were the remnants of the clouds that rolled through earlier. Far to the south, I could see the rock formations emerging from the beginning of the Badlands. In the distance, a few pumpjacks drew oil up from the ground.

If I squinted and let the power lines and pumpjacks go blurry in my vision, it wouldn't have been much different from the view my great-grandmother had in 1906. Certainly it smelled the same after a storm. The insects would have been just as loud, the sunflowers just as bright. The skies would have been just as dramatic, the wide, blue horizons framing castle-like white clouds.

A few hours earlier, I'd set out in search of Anna in the cheapest car I could rent, guided by little more than instinct and coordinates on the map splayed across the passenger seat. A harried attendant had waved me toward a dirty, low-slung Toyota in an overflow rental car lot, a muddy field next to the single terminal at the Minot airport. The airport had the capacity—just barely—to handle daily flights to oil industry hubs in Denver and Houston. No one had bothered to vacuum the Toyota's carpet or to wash its exterior. When people could make $100,000 per year or more driving trucks in the oil fields, it was challenging to fill jobs at fast food restaurants, hotels, and rental car counters.[1] In Williston, the smaller, dustier town at the heart of the oil fields 125 miles to the southwest, one McDonald's franchise offered workers $300 signing bonuses.[2] Some fast food restaurants paid as much as $15 an hour, more than double North Dakota's minimum wage at the time.

As I exited the Minot airport, I felt both exhilarated and terrified. It was one of the things I loved about my job, showing up somewhere new to me and working to understand its context as quickly as possible. I liked having a puzzle to solve, on deadline. I needed to figure out fast who to talk to, where to go, and what the story was without resorting to cliché or stereotype. I couldn't fail; otherwise, I was just another parachuting journalist who didn't quite get the full measure of a place. We seldom get the full measure of a place, of course, even when we live there. But I loved the effort of it and the excitement of arriving somewhere new to me. I got to justify my curiosity with my profession.

This time, though, I was driving into unfamiliar personal terrain. Three years had passed since my sister and I sat at my father's kitchen table, sorting out our mother's jewelry and shuffling through the oil company paperwork. I hadn't forgotten about Anna or the mineral rights in North Dakota, not at all. But so much had happened after my mother's death. My grandmother died on my father's side, unexpectedly leaving each of her six grandchildren a $19,000 inheritance. I used my windfall to pay off debt, including the $1,000 I owed my sister for loaning me the deposit on my last apartment.

I met Chris, someone I wanted to spend the rest of my life with. We moved in together and got married, easing some of the financial pressures of single life. Because I was in my late thirties, I turned my attention toward having a baby as quickly as I could. Every month, I tested my urine for the hormonal surge that indicated when I was ovulating. We were newlyweds, so all the timed sex was still fun. The pregnancy tests with their monthly no were not.

My job shifted, too. After President Barack Obama's reelection, I started writing about environmental issues and climate change for the Washington bureau of McClatchy newspapers. It was a beat full of potential, especially since the Obama administration had recently announced a climate change plan that had the possibility of transforming the nation's power grid by shifting it away from coal.[3] The president was making good on the pledge in his second inaugural address to respond to climate change. At the time, he cast it as a moral obligation. Failing to take action "would betray our children and future generations," Obama said in his speech.[4] A few months later during a speech in Germany, he called climate change "the global threat of our time."[5]

This came even as oil companies were perfecting hydraulic fracturing, the technology to push deeper into the earth to break up underground formations to extract oil and gas.[6] An oil and natural gas fracking[7] boom was underway in North Dakota, Montana, Wyoming, Colorado, New Mexico, Texas, Oklahoma, and Pennsylvania, injecting life into the postrecession economies of those places.

But the grand rhetoric about acting on climate change for "all posterity"[8] belied an ugly reality: Obama needed the energy boom to fuel economic recovery. The nation was still clawing its way out of the same recession that peaked the year my mother was dying. Fracking wasn't going anywhere, not yet. "The bottom line is natural gas is creating jobs," Obama said, even as he announced his plan to fight climate change.[9]

The mixed message both intrigued me and bothered me. As a reporter who wrote about political and environmental issues, I was always on the lookout for gaps between what people and companies and governments said they would do and what they were actually doing. It was clear that natural gas, still a vicious source of carbon pollution, was going to be a part of the energy mix, not just because it burned cleaner than coal, but because it was also crucial for a national economy still fragile from the recession. Hardcore environmentalists knew it: Obama wasn't going to save the planet.[10]

In my spare time, I tried to learn more about the North Dakota oil boom and the woman behind the mineral rights my mother passed on to us. I kept a file labeled "The Prairie Project." My great-grandmother, Anna, I learned, staked a claim to the land in 1905, at a time when many single women like her filed for homesteads in North Dakota. She got married soon after to my great-grandfather, Andrew Haraseth. So far, I knew little more about Anna than this.

I had hoped my employer would sign off on a reporting trip to North Dakota. There were plenty of environmental stories to write about the oil boom, and I figured I could tack on a few days of personal time to learn more about my family's mineral rights and Anna's history. But my editors said no. They weren't interested in paying for an expensive reporting trip out of an already minuscule travel budget. The company owned newspapers in Texas and Pennsylvania, and, not unreasonably, my editors suggested I focus my reporting efforts on the environmental effects of the oil and gas booms there.

It was clear that any research I did on North Dakota would have to be on my own dime, on my own initiative. So I cashed in some frequent flier miles and a week's worth of comp time and vacation days. I planned a weeklong trip to North Dakota in August to do my own research.

I had done it before. A decade earlier at the *Miami Herald*, an editor didn't want to pay to send me three hundred miles north to track down a corrupt official I had been writing about. The official was dodging my calls, but a source told me he would be at a folk festival near Gainesville, Florida. I complained to one of my more experienced colleagues, who told me to consider going anyway. "How bad do you want it?" she asked me. I wanted the story bad, I told her. So I paid out of my own pocket to drive north and spend the night in a hotel. I tracked the official down at the festival, interviewed him, and filed the story for Sunday's paper.

Since then, "How bad do you want it?" was what I asked myself whenever I weighed complicated choices that required personal or financial sacrifice and were accompanied by long odds of any return on my own investment.

I wanted to go to North Dakota. I wanted to go bad. I wanted to drive to the rectangular plot of land that seemed so obvious on the map, to see where Anna once lived. I wanted to find out what happened to her. I wanted to see this oil boom that I had only read about. Other journalists were writing about it, especially in the big, national magazines, and it felt as though I was missing out on an opportunity to make my own mark as a writer. I wanted to form my own conclusions about North Dakota and to write my own stories about the boom and climate change. And I wanted to find out more about my family's tiny role in it all.

Luckily, my self-financed trip coincided with a visit to North Dakota by then-Interior Secretary Sally Jewell, who was scheduled to tour a drilling rig actively producing oil on federal lands. Oil companies were secretive about their drilling operations, for competitive reasons, which made it difficult to get an insider's view. Jewell's visit was an unexpected opportunity to see fracking for myself, at a well owned by the legendary Oklahoma oil tycoon Harold Hamm.

I was also curious about how the Obama administration was balancing its ambitious climate goals with its fossil fuel agenda. Jewell seemed to embody these contradictions. She was a trained petroleum engineer who went on to manage the outdoor retailer REI. She had two impressive notches on her resume: she had both fracked a well *and* climbed Mt. Rainer in Washington state.[11] There was inherent conflict in her role, which mirrored my own ambivalence about the boom and the mineral rights my family inherited. I loved the idea of a windfall, but I didn't want to be a part of ruining the planet for my future children.

My editors declined my offer to write about Jewell's visit during my time off. You're on vacation, they told me. It didn't matter to me. Being a reporter often meant working irregular hours. Plenty of previous vacations had been undone by breaking news, with Labor Day weekend being a particularly unlucky working holiday for me and the people I wrote about. Labor Day 2005: Hurricane Katrina. Labor Day 2007: Larry Craig's arrest in a Minneapolis bathroom. Labor Day 2008: Sarah Palin's pick as the vice-presidential nominee.

On the surface, it seemed like they were caring bosses who didn't want me to work on vacation. The reality was more nuanced. McClatchy's declining fortunes meant everyone was overworked. There was often no one in the bureau to edit my stories, and the harried editors who remained often didn't have time to oversee my work. The newspapers I wrote for barely had space to run them. Climate change in 2013 was especially uninteresting to many of the local papers, I was told, especially if our stories predicted doom and gloom with no hope.

Hearing "no" all the time wore me down, and I worried about my career. I kept a running document on my computer of all the times my editors turned me down, just in case my productivity became an issue when it came time for my annual review. It was a discouraging place to be in, this constant state of rejection. I dreaded going to the office. It felt like my career was out of my hands, stifled by circumstances and economics out of my control.

I told the Interior Department press office I would cover Jewell's visit for McClatchy, and I signed up for the tour of production facilities in Williston. My press credentials gave me special access to important people I otherwise wouldn't have been able to meet. I felt a little bad about taking advantage of those credentials when I knew it was unlikely I would write anything about Jewell's visit for the people who paid my salary. But I wanted to see how fracking worked, and this was one of my best opportunities, even if I was technically on vacation. I held out hope my editors might change their minds if Jewell did or said something so compelling that they couldn't decline a story.

Really, though, I was there for my own curiosity. So I didn't care that I was running up my credit card again to pay for my own research. I was saying yes to myself and to the story I wanted to find, sidestepping the endless cycle of noes documented in my depressing computer file and the negative pregnancy tests.

I wanted this thing of my own so bad. It was something no one could say no to, except for me. I was going to find Anna.

———

En route to Anna's place.

Anna's old homestead was only one hundred miles away from the airport, but it took me all afternoon to arrive. I kept stopping to take photographs. Mere written description of the rural North Dakota landscape didn't seem to be enough. I needed a visual record of what I was seeing, to burn it into my brain.

I drove along Highway 52, through a valley carved by the shallow Des Lacs River, its meandering bends sometimes within just feet of one another. The drive parallels the railroad tracks that delivered Anna from Minot to her homestead in 1906. It was also a journey back eons, into the geological history of the glacial melt that had created the present-day landscape. Low, grass-covered hills, still green from a wet summer, sloped toward the narrow river plain.

At 1,900 feet above sea level, the valley felt to me like a system of foothills, not the open prairie I had expected. Physically and geographically, it was an interstitial zone. The Des Lacs dips down to North Dakota only briefly from Canada, before flowing eventually into Hudson Bay. Just a few miles away, tributaries of the Missouri River flow toward the Gulf of Mexico.

Flatbed trucks carrying drilling equipment barreled around me on a two-lane road built for tractors and slower-paced times. Tanker trucks carrying crude oil sped toward me. I could tell what was inside the tanks by the UN code, the diamond-shaped hazardous materials placard required by law. The number 1267 was everywhere: petroleum crude oil.

At a railroad crossing, I shut off my car's engine and waited for an oil train carrying crude oil to refineries. Hundreds of black tanker cars marked with UN 1267 snaked across the prairies on tracks built a century earlier to transport grain out of the state.

Halfway to Anna's property, I followed a sign toward Lostwood National Wildlife Refuge. The refuge, a 28,000-acre expanse of rolling hills, is one of the country's biggest publicly owned tracts of northern mixed-grass prairie. The wider ecosystem of the region is home to thousands of seasonal lakes known as prairie potholes, formed ten thousand years ago when glaciers melted and left depressions in the earth. In wet cycles, the potholes fill with snow melt and rain. In drier times, they slowly evaporate and drain into the water table below.

A one-hundred-foot fire tower, built in 1936 by the Works Progress Administration, overlooks the refuge. I climbed up to see an expanse of prairie as it must have looked in Anna's time, after the bison had been slaughtered to near extinction but before settlers had cultivated the land.

From the tower, the pothole lakes reflected back the sky: white clouds framed in blue. Anna, I imagined, stood on a ridge, framed by the never-ending, ever-changing sky. It was the height of summer and prairie roses were in full pink bloom. Everything seemed green and lush and full of promise. The wind rustled through the prairie grasses, a soft, ever-present background hum underscoring the dissonant, high-pitched buzz of insect life. The wind blew Anna's hat back to reveal her hair, a reddish-blondish-brown like mine. She had light freckles, like me. She rolled her shirtsleeves up, revealing forearms fierce from work.

Up there on the fire tower, I could hear toads burping below. Insects chirped. Waterfowl honked. And the wind blew, endlessly. It was unexpectedly loud for a place now empty of humans. In 1915, a scientist with the U.S. Biological Survey, Remington Kellogg, interviewed early homesteaders, who told him the ducks were once so cacophonous at night "they had to get up and shoot to scare them away so that they could sleep."[12] I was all alone up there in the tower on a Sunday, just me and a view of rolling green prairies. But this land had never been empty. Hundreds of thousands of bison once roamed these hills. Elk, deer, moose, and antelope remain even now. More than two hundred faded teepee rings dot the refuge, left behind most likely by the Assiniboine people who lived in the region before settlers like my great-grandmother arrived.[13]

Turning around, the view gave over to the man-made. I saw the muddy gash of a road being widened in the distance. I also saw what's known as a well pad, marked by silo-like water and petroleum holding tanks. There, I could see a pumpjack, the grasshopper-shaped symbol of the oil fields, bobbing up and down to draw oil from below the surface. I also saw a natural gas flare, a flame visible even in daylight as oil producers burned off the excess gas that came up with the oil.[14] Such flares, I knew, were a major source of methane and light pollution across North Dakota. From space at night, they made the sparsely populated oil fields look like a megacity. Invisible to the human eye were the colorless greenhouse gases entering the atmosphere, speeding up climate change.

I descended the fire tower and drove back through the refuge to the well I had seen from above. It was unfenced and unguarded, and I drove in to look closer. The flare hissed as it burned. It was so loud in person it canceled out the sound of birds and insects, even the sound of the ever-present wind. I photographed a sign, framed charmingly by wildflowers: "DANGER: Hydrogen Sulfide Poison Gas Present." I made a note to find out whether a woman who wants to have a baby is already too close when she sees the signs for hydrogen sulfide gas. I wasn't taking any chances, so I got back in the car and kept driving toward Anna's old place.

Along the way, I paused to photograph a faded red barn with a gambrel-shaped roof. The barn buckled precariously to the right, seemingly destined to collapse in the next big storm. And even later, I saw a homestead-era Lutheran church with its own off-kilter lean. I tried to photograph it, too, but the lean made it look as though I didn't know how to take a picture with a level horizon. *Did the wind knock everything askew?* I wondered. Or perhaps the wide horizons of the place disoriented everyone's perspective, my own included.

Technically, I knew where I was. My GPS showed me positioned in the far northwestern corner of North Dakota, so close to Canada that every ten minutes I got a text warning me in French and English about international roaming charges. I was in Burke County, population 1,968 in the most recent census. A few minutes earlier, I had turned left off County Road 5, and had driven south into Larson past a sign with tattered pennants advertising a long-closed supper club. It looked like it was once the most fun you could have had for miles and miles. There were few other signs of habitation, although several mowed lawns suggested the census takers found all twelve of the people who claimed to live there.

Once I crossed the railroad tracks and headed south on gravel roads and farm tracks, though, the orderly grid on the map in the passenger seat seemed to lose itself to the undulating prairie where my great-grandmother once staked a claim.

Anna's 160 acres should have been obvious out here. The outlines were apparent at home in Washington, DC, when I looked at the mapping software that the state of North Dakota used to identify oil wells. On the map, the boundaries of Anna's old homestead looked like a rectangle, set on its short end. The outline of her acreage was just as clear on the section line maps drawn up by the federal government. American mapmakers and surveyors have used this tidy system of townships and meridians since 1785.[15] The lines, championed by Thomas Jefferson, organized what the federal government thought of as wilderness into identifiable ownership parcels sold off to pay Revolutionary War debts. The coordinates rippled westward across the map from Ohio, in a linear quilt shaken out and draped over the continent, still in use as legal description of property even today.

I kept driving along what seemed like the boundaries of Anna's property. I was looking for archaeological evidence, like the dilapidated red barn I saw earlier in my drive. I imagined some sort of physical discovery to help answer my questions about Anna. In my mind, I saw a decaying old homestead shack, battered by time and wind and weather. It had a creaky door, hanging by just one rusty hinge. Sunlight illuminated dust motes in the interior through holes in the roof. Mouse droppings made me sneeze, which startled the birds roosting in a quiet, abandoned place. It was all in my head, but I imagined finding some clue, like an abandoned diary hidden beneath a floorboard. Nancy Drew meets *Little House on the Prairie*, my flawed but indelible childhood literary influences.

Yet when I arrived at the coordinates, all I could see was farmland mixed with uncultivated prairie. There seemed to be nothing out there to mark the place where Anna was a young woman. If anything existed, I couldn't identify it.

Any surveyor would have been able to find Anna's land; the oil companies knew exactly where it was. My mother, of course, would have known where we were. She drove to this place with her father and brother as a child, and had she been alive to accompany me, I know she could have figured it out, exactly. These were her people. This place was in her blood, even if she never lived in North Dakota. She raised two daughters to love wide open spaces, as I had written in her obituary. If I were to have a daughter, I wanted to teach her this, too. I would tell that little girl we were descended from a woman who made her own way on the North Dakota prairie—or so my mother told me.

My mother was never lost, at least in the geographic sense. Even when she took a wrong turn. When I was ten years old, my mother and one of her friends took me and a group of small-town fifth-grade girls to Portland, Oregon, handed us maps, and put us in charge of navigating the city. Everyone in Oregon went camping, she figured. Why not teach us skills we didn't know? We rode the bus, learning how to drop our coins into the fare box and how to puzzle out the routes that took us where we needed to be. We stayed in a hotel, some of the girls for the first time ever. We learned how to tip a then-generous 15 percent. It was from my mother that those other girls and I learned something all women must understand: if you know where you are, you always know where you stand.

Here I was, though, lost. And with a very different kind of life than that of my mother at the same age. There were no ten-year-old girls looking to me for guidance or inspiration. After months of negative pregnancy tests, I was beginning to doubt I would ever even have a ten-year-old of my own.

Before I left Larson, I took some photos and made a few notes. I stopped near a small grove of eastward-leaning cottonwoods on an otherwise treeless stretch of road that I thought paralleled Anna's property line. I wasn't sure I was photographing the right spot, but something about it called to me, something deep in the cells of my bones. Someone planted the trees in the past, perhaps Anna or her husband. Perhaps someone who once knew them. Planting a tree requires an eye toward the future. Not big grand thoughts about the march of time and our insignificance in the face of all of it. But the wisdom that someday, given enough time in an unforgiving place, the trees would grow into a small refuge from westerly winds.

I drove back toward Minot, headed for a residential hotel where I had rented a kitchenette. It cost $500 for the week, a relative bargain even on the edge of the oil patch, and one made possible only because it was so new it wasn't yet listed on all the travel booking websites.

As I drove, I found myself lost in uncertainty. It had seemed so simple when I planned my research trip. Fly to Minot. Drive to the coordinates on the map. Go to the scene, a fundamental tenet of journalism. See what you see. It felt like a failure, not knowing the exact physical location of Anna's homestead. If I couldn't find a place that was so clearly outlined on the map, maybe I wouldn't be able to find anything about Anna, about these mineral rights. I wondered if my instincts were off, if all the "no" from editors was because I wasn't a good reporter or writer. Maybe there was no story. Maybe all the time I'd spent writing about other people's stories was swirling around in my brain, manufacturing a sense of mystery and suspense where there was none.

Yet I also knew I couldn't face the possibility of another no. I hadn't yet made any promises to Anna, but telling her story seemed like a promise to my mother and to myself that I wanted to honor. I had a week away from the office to figure this out. I knew I wanted it too bad to give up.

Perhaps all journeys start out this way. We think we know where we're going, what we're looking for, what we'll find when we get there. It's not until much later that we understand: we have no idea what we're looking for when we first start out.

The days are long in North Dakota in the summer, but it was dark when I arrived at my hotel in Minot. I went to bed thinking about the sky. It dwarfs everything, even in places where there is the visual relief of buildings, signs, cars, and people. I dreamed of towering white clouds so vivid they were almost part of the landscape themselves.

I had other scenes to see in the morning, under those skies.

3

RIGGED

August 2013

North Dakota crude: *$97.18 per barrel*

"Hey all, what are the logistics for tomorrow?" I asked in an email, my nervousness clashing with my excitement to see fracking firsthand. And would I need to bring my own steel-toed shoes? Safety glasses? Or a hard hat?

No, the Interior Department's press office reassured me. The oil companies would have protective gear and a safety briefing for reporters who were touring the drilling operations in Williston. They also sent turn-by-turn directions, which was a relief after getting lost trying to find Anna's place. Bulldozers were carving new roads daily into the grasslands and Badlands of western North Dakota. Most of the tracks were gravel and led only to drilling pads, but they were so new they didn't register on GPS or Google Maps. As I had already learned when I almost got mired in the mud near Anna's, it was best to know where you were going before you got on the road in North Dakota.

I left for Williston early, the sunrise in my rearview mirror as I drove due west. Highway 2 was a 125-mile straight line through the heart of North Dakota's first oil boom region in the 1950s, pointed directly toward the next big thing. Wheat fields lined the road, interrupted by pumpjacks, flares, and man camps, the temporary housing for oil-field workers. New mingled with relics of past boom and bust cycles—crooked prairie churches and weary barns and windmills from the homesteading era alongside the rusty pumpjacks that have been sucking up oil since the first boom.

Although now four lanes wide in most places, Highway 2 was essentially the same route driven sixty-two years earlier by another journalist, a photographer for the *Williston Press-Graphic* named Bill Shemorry.[1]

By 1951, wildcatters had been trying to drill for oil in North Dakota for decades. The geology seemed ideal, but the harsh winters and the vast distances to refineries held back development, as did the Great Depression and price controls on crude oil during World War II.

But in the winter of 1951, the state thrummed with the possibility of an oil boom.[2] Everyone in North Dakota was curious about the oil derrick rising above Clarence Iverson's wheat field just south of Tioga, a farm town about eighty miles west of Minot and fifty miles east of Williston. Crews with Amerada Petroleum Corporation had been drilling their way more than eleven thousand feet into the earth.[3] If Amerada struck oil, it meant there was oil in North Dakota. It meant a lot of people in North Dakota were going to be very rich.

On the evening of April 4, 1951, Shemorry got a call from his editor. Grab your waders and your camera, his boss told him. The Iverson well is coming in.

It was dark, but Shemorry could see the well long before he arrived. A worker had tossed a flaming, oil-soaked rag into the natural gas venting from the well. The gas caught fire, shooting thirty-foot flames so high and bright they were visible as far as ten miles away on the dark prairie. The intensity drew onlookers from miles around. Cars crowded the shoulders of the road, so many that as Shemorry drew near, he could barely find anywhere to park. The burning gas roared with such volume that people in the crowd had to shout to be heard over the noise.

The ground was still frozen. Melting snow with nowhere to go puddled on roads and low places. Near the flare, the heat began thawing the soil. Steam rose from the earth. Shemorry waded into what turned out to be a picture-perfect reflecting pool for his soon-to-be-iconic photo. Light from the burning gas flare illuminated the drilling rig, setting it off against the dark night sky. It was so bright that even in the dark of night, Shemorry could read the aperture and shutter settings on his 4x5 Speed Graphic, one of the boxy cameras used by professional newspaper photographers in the 1950s.

Shemorry took about a dozen photos. It was a true scoop. No other press photographers were there to capture a photo of the first commercially successful oil well in North Dakota.

In a rush to meet his deadline, Shemorry drove to the *Minot Daily News* to process his negatives in the darkroom. Then, he bought a sandwich and drank a beer while he waited for a photoengraver to make the printing plate used to press the image onto the front pages of the *Press-Graphic*. It was after midnight when Shemorry finally drove the printing plate 125 miles west back to Williston over flooded roads. He drank a celebratory six-pack of beer along the way, tossing the empties into the back of the car as he drove.

Local papers eagerly reprinted Shemorry's photo.[4] So did *Life* magazine and *U.S. News and World Report*. Shemorry's photo from Iverson's wheat field was vivid proof to the world that there was oil in North Dakota. The American Petroleum Institute took out ads in national magazines proclaiming North Dakota "the new oil state."[5] Almost overnight, every farmer in the Williston Basin began dreaming of black gold beneath his wheat fields.

For Alfred Jacobsen, the chief executive of the North American division of Amerada, the photograph proved what he understood was possible, with enough technology, investment, and determination in a harsh climate. Geologists at Amerada, a British oil company known for its early and successful embrace of subsurface seismic technology,[6] were convinced there was oil beneath the prairies of northwest North Dakota. Amerada bought a block of North Dakota leases once owned by Standard Oil, which had tried unsuccessfully to drill in the Williston Basin before World War II.

After the war, Amerada's geologists spent four years in North Dakota mapping the surface and underground formations and buying up leases.[7] By 1952, the year after Shemorry's photo was seen around the world, Amerada held 1.5 million acres in leases in the Williston Basin. It was the biggest block of leases held by any oil company then drilling in the state. In December 1952, Jacobsen made a prophetic prediction to *Time* magazine: "The Williston Basin is not just one oil-field. It is an oil province."[8]

Danish by birth, Jacobsen worked in the Mexican oil fields in the 1920s before joining Amerada. *Time* described him as a "tall, spare man with a lined, ascetic face, bright brown eyes and explosive energy." American oilmen, the article noted, called Jacobsen "the Great Dane." Jacobsen lived with his wife in an apartment at the Plaza Hotel in New York City and commuted to work on the subway. He was a workaholic with almost no social life who devoted most of his time to making money.

"I'm not gregarious," he told the magazine. "I have business associates, but no personal friends. I don't go in for that sort of thing." He told *Time* that wildcatting for oil was the "easiest thing in the world." You can make millions and millions, he promised. "All you need is a checkbook—and money in the bank." He let the magazine in on another secret: oil companies were allowed to write off the entire cost of drilling a dry hole. The depletion allowance granted to oil companies in the 1920s continued to let them deduct 27.5 percent of their gross revenues. Oilmen of the era didn't just require skilled geologists. They needed lobbyists to persuade lawmakers to write friendly legislation and accountants who knew all the loopholes.

I puzzled over the midcentury obsession with oil tycoons, even as I drove toward Williston to see the 2013 version. Jacobsen was a more taciturn type of the flamboyant Texas wildcatters that *Time* kept featuring on its covers. The editors were especially enthralled and repelled by Glenn McCarthy, the man who inspired *Giant*, the 1952 novel by Edna Ferber that got made into a 1956 film starring Rock Hudson, Elizabeth Taylor, and James Dean. McCarthy even looked the part, *Time* wrote in a 1950 cover story. "After 17 spectacular years in the Texas oil fields, he looks like nothing so much as a Hollywood version of a Mississippi River gambler," the reporter wrote, "a moody and monolithic male with a dark, Civil War mustache, a cold and acquisitive eye, and a brawler's shoulder-swinging walk."[9]

These archetypes of 1950s masculinity were fixtures of Hollywood storytelling, too, which made sense. The film industry's origins in California paralleled the state's own rising oil industry in the early 1900s, for one. But the oil fields also proved rich source material for emotionally compelling stories with larger-than-life main characters caught up in a big, unanswered narrative question that preoccupied most Americans: Would they strike it rich?

An oil field, as Shemorry proved with his photo and as I could see as I drove, also made for cinematic visuals. There were gushing oil wells, the oil derricks, and pumpjacks silhouetted against wide skies. Then there was the physical industriousness of the brawny oil field workers themselves. Clark Gable, who had in fact as a teenager worked as a roustabout in the Oklahoma oil fields, in 1940 cut a dashing figure in *Boom Town*, an oil patch film that celebrated the individual right to pursue riches no matter the cost to the environment or to the people trampled in the path of progress.[10]

By emphasizing the romance and adventure and pluck of wild-catting, Hollywood made it seem as though anyone could get lucky. It was merely the next evolution of the myth of the rugged pioneer out to conquer untamed wilderness.

But the truth was much uglier. Jacobsen hadn't even tried to deny it in his *Time* interview, when he cut to the heart of capitalism: to make money, you had to *already* have a checkbook full of it to help you rig the system in your favor. The system was corrupt, and I knew it; I'd seen its effects on my own family, in that ledger my mother kept of her medical debts. Muckraking journalists like Ida Tarbell knew it, too, more than a century ago. In 1904, Tarbell published a trust-busting account of Standard Oil that led to the breakup of the company founded by John D. Rockefeller, the first oil tycoon and the richest man in American history.

"I never had an animus against their size and wealth, never objected to their corporate form," Tarbell wrote of the company in her 1939 autobiography.[11] "But they had never played fair, and that ruined their greatness for me."

In a winner-take-all system, these men were accountable only to their own fortunes. They got rich by plundering the earth, by taking advantage of lax environmental regulations that prioritized short-term extractive techniques over stewardship, and by strong-arming lawmakers to enact generous tax breaks that allowed oil companies to write off the cost of exploration.

But much like the muckraking Tarbell and the midcentury journalists at *Time*, I still found myself fascinated with these tycoons. It took a certain amount of leading man energy to believe the spoils of the earth were yours for the taking. These sorts of men—and they were always men—didn't take no for an answer. They did what they wanted, which not only took tremendous self-regard, but was also likely at the heart of their masculine appeal. They didn't ask for permission to pursue their dreams. They had an individualistic swagger that verged on repellent, but could come off as dashing or heroic if you didn't look too closely at why we as Americans allowed them to exploit the law, the earth, and our fellow working people.

I turned down the road leading to the Continental Resources facility in Williston and parked my dirty rental Toyota. I was eager to see firsthand the modern manifestation of an oil tycoon: Hamm, the Oklahoma billionaire who'd bet big on North Dakota.

Inside the warehouse, the safety officer handed out blue smocks and white hard hats to the reporters and visitors who were going on the tour. I finally got an answer from him to my question about the danger of hydrogen sulfide gas: it was, in fact, a very big deal. Women like me who wanted to get pregnant weren't the only ones who needed to pay attention to the warning signs like the one I'd seen near the wildlife refuge. Many workers wore hydrogen sulfide sensors on their safety vests to alert them to unsafe levels of the rotten egg–smelling gas.[12] Several workers had died from exposure at the height of the boom, one of many grim facts that led the AFL-CIO labor union to designate North Dakota for a few years running the most dangerous state per capita for workers.[13]

After the briefing, I finally got a look at Hamm for myself. Ruddy and round-faced, he was a relatively tall man with an above-the-buckle gut. He wore jeans, a black Continental windbreaker, and one of the same white hardhats the safety officer gave me. Hamm, who picked cotton as a child[14] before getting his start as a truck driver in the Oklahoma and Texas oil fields, looked like an ordinary sixty-seven-year-old white man who had once made a living with his physical labor. He was, I realized, the same age as my father. They even looked a little alike, with their round faces and faded reddish hair. Hamm had almost certainly flown to Williston on a private jet. But other than the deference shown to Hamm by all the elected officials and employees at the facility he owned, he had none of the trappings of the oil tycoons of lore.

Hamm's stake in Continental in 2013 was said to be worth $11.3 billion,[15] and Hamm himself was ranked No. 90 on the *Forbes* list of richest people in the world. He was also in the midst of a messy divorce, a $1 billion settlement that made it one of the most expensive splits in U.S. history.[16] In one profile, Hamm said he had a "gut feeling" about North Dakota as his company and others were perfecting horizontal drilling techniques.[17] He began snapping up then-cheap mineral rights across North Dakota. The headline in yet another profile described Hamm as "the man who bought North Dakota."[18]

And indeed, Hamm had an outsized influence in North Dakota. In 2012, he gave $5 million personally to the geology program at the University of North Dakota and his company gave another $5 million.[19] His political donations fueled campaigns for statewide and federal office in North Dakota: Hamm gave $20,000 to then-Governor Jack Dalrymple's 2012 campaign.[20] The same year, Hamm served as an energy adviser to Utah Senator Mitt Romney's presidential campaign.[21]

Hamm did not have the same influence in 2013 with the Obama administration, which was developing fracking rules for federal lands, including how much gas could be flared from wells.[22] It's why the U.S. Interior Secretary, Sally Jewell, was also there on the tour, to see in person what was going on as the agency finalized regulations around flaring.[23]

Another iconic photo, like Shemorry's in 1951, was influencing public opinion around fracking. A few months before I arrived in North Dakota in 2013, NASA released images from a new satellite. Launched in 2011, the Suomi National Polar-orbiting Partnership satellite had the expanded capability to capture more precise images of what the planet looked like at night from space. The satellite's sensors could capture tiny sources of light, like the beacons on ships in the middle of the ocean and the glow of street lights on dim country roads. NASA boasted it was an unprecedented "global view of the human footprint on the Earth."[24]

One blotch of light on the satellite images caught everyone by surprise. From far above the earth, the satellite images made it look as though there were a megacity on the plains of North Dakota, one of the most sparsely populated places in the country.[25] The source of light was quickly identified: the Bakken oil fields of North Dakota.[26] The satellite imagery captured lights on drilling rigs and other infrastructure of the oil boom, along with the light produced by methane flaring from the new oil wells.[27] The North Dakota oil fields had a new nickname: "Kuwait on the Prairie." (In 2015 a North Dakota state agency disputed how the images were processed, saying it gave "the uninformed public the idea that flares are literally lighting up many square miles of prairie countryside.")[28]

There on the ground in Williston, I was seeing firsthand what produced all that light. From Continental's twenty-acre drilling site, oil workers bored two miles down and two miles horizontally. There were fourteen wells at the time of the tour, all drilled down and then out, in multiple horizontal directions. Drillers could even tunnel under the Missouri River and Lake Sakakawea, which seemed an obvious risk for water quality.

Unlike the big North Dakota boom of the 1950s, oil companies weren't just tapping vertically into underground reservoirs of petro-leum. Instead, they drilled down deeper into harder rock formations. Then they drilled horizontally along those formations, accessing their full breadth. Once the drilling was complete, crews fractured the for-mations with a high-pressure mixture of water, sand, and proprietary chemicals to free the oil and gas trapped between rocks. The fracturing was why it was called "fracking."[29]

Continental boasted of its innovative work developing hydraulic fracturing techniques in North Dakota, beginning with the first hor-izontal well the company drilled in 2004 near the town of Crosby.[30] When I looked at its location on a map, I was surprised to see that the company's first fracked well in North Dakota was just fifteen miles from Anna's homestead.

As proud as they said they were of what they were doing,[31] the company didn't want us taking any photos of the Atlanta drilling pad, the area the company used to stage some operations in Williston. Continental sought to persuade Jewell that their physical footprint here was slight. But she wasn't fooled.

Jewell told me several years later that she'd experienced something similar in Wyoming, where oil companies opposed efforts to list sage grouse as an endangered species.[32] She'd toured a drilling pad where the oil producer touted its minimal footprints. She asked the company to show her pilot the coordinates of one of their new facilities so they could fly over it on their way out of town.

"They got really nervous about not wanting to show me where it was," she told me. "And I tell you, it was a hell of a lot more disturbed than what they were painting a picture of. The same is true in flying over lots of areas that otherwise are so remote nobody goes there. You fly over and you look down and you think, 'oh my God, I know what I'm looking at. This is just devastating.'"

At the Continental pad in Williston, I complied with the photo prohibition, but reluctantly. I told myself I would have made more of a fuss about the photo ban were I actually writing about Jewell's visit for my newspaper company. But I also knew we were allowed to take as many photos as we wanted at our next spot, a drilling pad owned by the Norwegian oil company Statoil. I figured I'd get what I needed there.

As the Continental tour commenced, I traipsed behind Hamm and Jewell, eager to observe their interactions. In person, Jewell was complimentary and polite, but not as deferential as the rest of the officials at the gathering, including the state's two senators. Both Heidi Heitkamp, a Democrat, and John Hoeven, a Republican, tripped over themselves to talk about how North Dakota would soon be pumping out a million barrels of oil each day. Both senators were eager to assure Hamm they believed in the Bakken boom, and that he had their full support. Their quotes were so anodyne I barely bothered to write them down in my notebook.[33]

Jewell had the background and authority to hold her own with Hamm, and unlike the elected officials on the tour, she didn't need to worry about her next turn at the ballot. She, too, had fracked a well on one of her first jobs out of college, as an engineer for Mobil. After that, she moved on to banking, where she evaluated oil and gas investments for their soundness. Eventually, she led REI and sat on the board of the National Parks Conservation Association. Throughout her career, Jewell used her training as an engineer to understand how systems worked. She told me she thought it helped her see beyond the trappings of power when she met executives like Hamm or Rex Tillerson, the former CEO of Exxon.

"I have both an engineer's and a banker's brain, which means that there's an element of skepticism," Jewell said. "I'm not just going to buy what people tell me. I think that the desire to really understand how things work made me a little less susceptible to hype and the intimidation that comes with power."

Hamm was irked he had to defend what he was doing. "It would be very, very good if we didn't have a government that was intent on putting us out of business," he told a reporter with *National Journal,* in the days following the tour.[34] "They don't like oil and gas. They don't like fossil fuels. They want to end us."

At the time, oil companies in North Dakota were flaring about 30 percent of the gas coming out of new oil wells. (Hamm estimated on the tour that Continental flared about 10 percent of its gas.) It was a shockingly high number compared to other oil-producing states, and one that federal energy regulators found alarming. State regulators gave oil producers a year to get new wells under control, and then in theory, the oil companies were required to pay royalties and taxes on the gas they were burning off as waste. But there were many ways to get a waiver.[35]

Some flaring is inevitable with oil and gas production; volatile gases can accumulate and must be burned off to avoid spontaneous combustion. That's why oil refineries have flares, for example. But North Dakota in the early days of the Bakken oil boom lacked the capture or pipeline infrastructure to produce all of the natural gas coming out of the wells. The gas was considered an unusable byproduct. Oil companies had little incentive to do anything but drill for oil as fast as possible. Oil made them far more money in North Dakota, and they were in a hurry to get rich. So they flared the gas as waste, burning it off at the source.

In other parts of the country, especially Pennsylvania, oil and gas companies actively fracked wells for natural gas. Its production was among the reasons coal plants were shuttering across the country. Gas was cheaper and less obviously destructive than coal. Unlike the sooty plumes emitted from coal smokestacks, the emissions from a natural gas plant weren't visible to the naked eye. But methane, the main component of natural gas, is more than twenty-five times as potent as carbon dioxide at trapping heat in the atmosphere.[36] It is also not as long-lasting of a greenhouse gas—it begins breaking down in the atmosphere in a matter of decades—so slashing methane emissions is seen as a quick way of slowing climate change.

Flaring all that gas just to access another fossil fuel was a massive plunder of resources. Earlier that summer, I'd been struck by something I read in a report by Ceres, a research organization that encourages sustainable investment. It found that the flaring in North Dakota was equivalent to the annual greenhouse gas emissions of adding a million cars on the road.[37] Beyond the climate degradation, it was a wasteful practice. Thousands of homes could be heated with all that burned-off gas. "People are estimating it's about $1 million a day just being thrown into the air," Marcus Stewart, an energy analyst with Bentek Energy, told NPR. "It's significant because we don't really see anything like that anywhere else in the country."[38]

I looked to the bluff above the Continental drilling pad. There sat a brand-new housing development, built just south of town to accommodate the influx of people to Williston. The drilling pad I was visiting was so close to the subdivision that Continental Resources was actually a member of the homeowner's association.

The nearness of the housing development made the photo ban even more exasperating. If anyone in those homes could see everything happening at the facility, what was the purpose of keeping us from taking pictures? I hoped no one from Continental was responsible for enforcing the rules of the homeowner's association. I could picture the oil company representative as one of those obnoxious neighbors eager to call the city code inspection office when your grass grew too tall.

I also worried about the people who lived in the development. Even under the best of circumstances, there was a growing body of scientific evidence showing that living next to any sort of oil production facility exposes people to volatile organic compounds that cause breathing problems and possibly even cancer.[39] Yet housing was scarce in Williston, and few of the newcomers had many options. Living next to a drilling pad wasn't really a choice. It was one of the many short-term compromises people resigned themselves to in exchange for jobs and stability that might outlast the boom and bust cycle. Even if their housing threatened their long-term health.

The dog at the Statoil rig in Williston, North Dakota.

That afternoon, Jewell toured another facility a few miles away, one owned by Statoil. Statoil was just as strict about safety but more open about its operations. I noticed a border collie hanging out on the drilling rig as we climbed up in our blue protective coveralls. Like the humans working on the well, the black-and-white dog was covered with lubricants and dirt and other muck. To get to the drilling platform, the dog had to climb up about three flights of yellow, metal steps. She was unafraid to do so, and way too friendly.

"Mascot?" I joked with one of the roughnecks.

"Rig dog," he said, grinning.

I snapped a photo. The workers told me the dog was from a nearby farm. She kept coming over to see the people at work, even after the workers repeatedly shooed her away. It was the kind of dumb thing my own overly friendly black-and-white dog would have done when she was younger and liked digging under fences to roam the neighborhood. I found myself worrying about the rig dog's exposure to low-to-the-ground hydrogen sulfide gas and whatever else was in the gunk soaking its fur. Mostly, though, the rig dog's presence showed how it and many other facilities were open and unfenced, accessible to anyone who passed by, including curious dogs.

Jewell and her team at the Interior Department visited Statoil because the company was developing mobile condensers, units that could be delivered to oil wells to gather natural gas until it could be piped away for processing. Her visit was a loud signal to the industry that the flaring free-for-all was over. Even if she didn't say it in person, it was clear in the press release her communications staff released after her visit to the oil fields, and then later in the day, at her trip to Theodore Roosevelt National Park on the western side of North Dakota.[40]

She wore her two hats uneasily: drilling boss and chief executive of the National Park Service. It was just as awkward in the press releases about her visit. As drilling methods and technologies advance, Jewell said, energy production has to happen "in a safe and responsible way for the environment and for communities." But the landscape of the park is also "a powerful reminder that, even as we bear witness to a production boom in the Bakken, there are places important to

America that are too special to drill and must be protected for future generations."

The contradictions unsettled me. Maybe it was time for the Obama administration to pick sides. A lot was at stake, including the future of the planet. And yet I wasn't any better; I hadn't picked sides yet, either. If I was truly honest with myself, I wasn't just in North Dakota to find out what happened to my great-grandmother, Anna. I was also there to find out whether there was even the tiniest possibility that someone would strike oil on my family's land. I, too, could hear the whispers. And I was beginning to understand that I was just as susceptible to being drawn to fortune, even as I found myself repelled by the consequences of greed.

It was late afternoon when the tours ended, and I drove to The Williston, a steakhouse in a Tudor-style former Elks Lodge. One of the public relations executives for Statoil told me I ought to eat there, if only to experience the high roller side of the boom. It was, he said, the swankiest place in town.

I was too early to see the action at full tilt, so I sat alone in the predinner hush of the dark interior, allowing the events of the day to sink in. I'd seen firsthand under big open skies how a well is fracked, my main goal for the day. From the tour, I understood the toll of fracking on the land, on the health of the people who lived nearby. There was no doubt flaring revved up the greenhouse gas emissions contributing to climate change. I took note of the mud on the faces of the roughnecks working the rigs and the clean fingernails of the public relations executives. I watched the way two U.S. senators acted when they were desperate to keep the spigot on in their state. I witnessed an oil tycoon try and fail to persuade a cabinet official to ease the rules so he and others could get richer by flaring more gas. I'd seen not just the mechanics of extraction, but the inner workings of political influence. And I had a connection to it all, merely because of a great-grandmother who, more than a century earlier, staked a claim on the prairies.

Earlier that day, I ran into Amy Harder, a reporter I knew from Washington, DC, who wrote about energy for *National Journal*.[41] She also was in North Dakota for the tour with Jewell. That morning, she had interviewed Hamm. Harder told me they ate biscuits and gravy together at the restaurant in the Holiday Inn Express where she was staying. He was chatty about his prescient role in identifying the promise of North Dakota, she told me, but close-lipped about his pending divorce back home in Oklahoma.

I envied her instinct to seek out a breakfast meeting with a billionaire.[42] It didn't occur to me to even ask for an interview with Hamm, mostly because my editors made it clear they weren't interested in stories out of my trip to North Dakota. The self-censorship scared me. I no longer thought creatively because I knew my bosses would turn down my ideas. Not only was I missing opportunities, but also the chorus of no was dulling my instincts.

Here I was in North Dakota, eager to write about what I'd seen, and it felt like no one wanted it. Like Shemorry, I was a grab-my-waders-and-go kind of reporter. But it felt like no one I worked for cared if I went anywhere. It felt as though all the "no" was chipping away at the essence of who I was as a journalist and a creative human.

If I looked at the Bakken oil boom in the most charitable light, it represented people who were unwilling to take no for an answer. They found more inventive ways each day to make money. Maybe there were some lessons in how they approached their work, how everyone from strippers to roustabouts swarmed the state to seek their fortunes before it all went bust. Everyone in North Dakota was out to make their cut. Maybe, like me, they were tired of hearing no all the time.

From what I'd read, Hamm didn't seem especially introspective.

Few millionaires, let alone billionaires, bother with much self-reflection. It gets in the way of making money. But maybe no one had asked Hamm the right questions; maybe they were too cowed by the power of his wealth. Maybe we all needed to summon up the swagger of the *Time* reporters writing profiles of midcentury tycoons.

If I had the chance to interview Hamm over a steak at the Williston, I knew exactly what I wanted to ask: the kind of questions that would get me kicked out of the place. I wanted to know when he first understood the world was his for the taking. How did he learn that the rules weren't for him or anyone like him? And I wanted to ask him why he thought he deserved to amass an $11.3 billion fortune when people were living in their cars in a Walmart parking lot, and in church basements and in housing next to his oil fields, just so they'd have a crack at a few crumbs of the American dream.

I splurged on my dinner like the oil heiress I was not: a grass-fed buffalo rib eye that was the least expensive steak on the menu, and a glass of pinot noir. It came to $48, before my $10 tip. It was, I realized, exactly a day's earnings for someone making minimum wage in the rest of the country, where salaries weren't inflated by the oil boom.

4

THE BOOM

August 2013

North Dakota crude: *$97.18 per barrel*

WE COULD BE RICH.

It might not have been said aloud, but it's what I imagined was whispered all throughout my mother's 1950s childhood. The words were like the wind, they were everywhere. Anything was possible with hard work, American know-how, and determination. It was even better if you owned a wheat field in northwestern North Dakota, like my grandfather did. It was entirely possible an oil company might plunk down a rig and make you a fortune.

As oil companies began leasing wheat fields from North Dakota farmers in the early 1950s, my grandfather watched with great interest from his home in Montana. Anna's former homestead was fifty miles directly north of the Iverson farm in Tioga—the one where Bill Shemorry photographed the first successful oil well, the one that signaled with its flare the potential riches coming to the farmers of western North Dakota. It wouldn't have been unreasonable for my grandfather to dream of a windfall on the land he inherited from a mother he never knew, not after the hard times of his youth as an itinerant laborer and later as a soldier in the South Pacific. My mother and my uncle remember their father sitting in his recliner after work in the 1950s, paging through his newly arrived *Williston Basin Oil Review.* He was always certain his lucky day was coming.

Their father was among the thousands of young men who left North Dakota during the Great Depression, in desperate search of work. The state was grim in the 1930s, struck hard by the drought and economic collapse of the Dust Bowl years. "Here and there drifting soil covered fences, ditches, and farm machinery," wrote Elwyn B. Robinson in his history of the state.[1] People drifted away, too, just like the soil. The state lost nearly forty thousand people between 1930 and 1940; those who remained may have been too poor to leave. In 1933, the per capita income in the United States was $375; in North Dakota, it was only $145, Robinson wrote. In 1934, the driest year on record, the state estimated that as many as half of its people were on government relief of some sort. Times were so hard in northwestern North Dakota that it drew the attention of Russell Lee, one of the Farm Security Administration photographers who roamed rural America documenting the need for and the effect of federal New Deal aid programs. I've clicked through the online catalog of Lee's photos of Crosby, the nearest town of any substance to Anna's old place, wondering if I was related to any of the farmers he photographed on a bright, dusty Saturday in 1937. Lee's camera gazes with the same longing as his subjects, their eyes fixed on the unattainable goods in the storefront window displays.[2]

My grandfather didn't speak of the hard times. His children remember only that their father picked apples in Washington state and worked as a lumberjack in Montana during the Depression before settling in East Helena in the late 1930s. There, he worked as a laborer in the Asarco lead smelter, one of the biggest employers in the region.

The plant's five smokestacks and its massive slag pile defined the town's skyline and its culture for more than a century, until its closure in 2001.[3]

In 1941, my grandfather Ed was drafted into the U.S. Army. Before shipping out to the South Pacific, he married my grandmother, Irene, who already had two young girls from a previous marriage. During the war, Ed served as a sergeant with the 116th Medical Battalion in the Papua and New Guinea campaigns. The Army discharged my grandfather on August 31, 1945.[4] He was thirty-nine. My mother was born exactly nine months later, among the first of the postwar baby boomers.[5]

Ed inherited the homestead from his father, Anna's husband, Andrew, who died shortly after his son returned from combat. My grandfather never lived in North Dakota again, but over the years, he leased the fields he owned to other farmers. He also got money from a U.S. Department of Agriculture conservation program established during the Dust Bowl years to keep the land out of production.

On May 22, 1951, my grandparents signed the first oil and gas lease on Anna's old land. Ed was a school custodian at the time. He and his wife liked piling the kids and their fishing poles in the car for a Sunday drive. He was not an exuberant or extravagant man, but he and my grandmother must have signed the lease with such hope and such anticipation. Perhaps Ed picked my six-year-old mother up and twirled her around with excitement, gleeful with the promise of what he could give his young daughter, if only they struck it rich.

———

Sixty-two years later, I drove to the tiny town of Bowbells to unearth the original lease at the Register of Deeds office in the Burke County courthouse, a three-story brick building with understated art deco lines and a listing on the National Register of Historic Places.

The Burke County Courthouse in Bowbells, North Dakota.

Bowbells was quiet at midday, even in the midst of the oil boom that overtook much of the rest of western North Dakota. I passed a charming purple cottage and wondered what it was like to live in such an unhurried place. I stopped the car and stood in the middle of the street taking photos of the house. There was no traffic. Nearby, a child left a bike abandoned on a front lawn, certain no one would steal it.

The courthouse was the second biggest place in town, after only the grain elevator. It was noteworthy mostly for who designed it: Toltz, King and Day, the St. Paul, Minnesota architects behind the Prince of Wales Hotel, a timber-framed chalet at Waterton Lakes National Park in southern Alberta, Canada. Hiring a well-known firm in 1928 to build the courthouse was a brazen act of pre-Depression optimism for such a small, remote place. But in 1920, there were ninety-five hundred people in Burke County, almost five times as many as its present-day population of twenty-one hundred. Its first white settlers saw themselves on an upward trajectory and they hired a fancy architectural firm from the nearest big city to build them something to stand the test of time.[6]

Nearly a century later, the Burke County courthouse remained a solid, brick presence. Its exterior seemed little changed from the 1920s, other than the patina from thousands of hands on the walnut handrails of the stairs leading to the main floor. Directly inside the courthouse on the wall to the left was a framed portrait of George Washington; to the right, Abraham Lincoln. Below Lincoln was a twelve-year-old poster commemorating September 11, 2001, printed with a flag and a message: "God Bless America."

I turned toward the Register of Deeds office, set behind frosted double-pane windows that the clerks shut firm for a half hour at lunchtime, and then again at the close of the business day. One of the clerks rose to greet me at the marble counter.

Some clerks reflexively deny access, both to the public and to journalists. Some are uninterested in why you're there, seeing you as a bother to their routine. They sigh when they're forced to press away from their tasks, but they'll grudgingly show you around, especially if you admit you have no idea how to find what you're looking for. And then there are those who will bend over backward to help you with your search, but you might have to share your life story in exchange for their assistance.

I could tell immediately by the candy dish on the counter: in Burke County, they enjoyed visitors and would gladly help. But I would have to explain my connection to the place and how I was looking for traces of a great-grandmother lost to history, never even known by her own son.

The courthouse was quiet, and the few other people in the lobby could overhear me telling the clerk what I was looking for. One man, intrigued by my conversation, came over and offered to help me find

Anna's records. My search was the most interesting thing happening in the courthouse that day, he told me.

The eavesdropper worked as a landman, one of the white collar jobs associated with oil booms. A combination of title search expert and salesman, a landman takes care of the paperwork after oil company geologists determine where they want to drill. It's the job of a landman to head to the county courthouse to conduct title and mineral rights searches on the land in question. To do this, they search all real estate transactions, often all the way back to the original homestead paperwork. Some of this is digitized, but many older records are not online, or if they are, the scans are not legible enough to read clearly.

In the early days of the Bakken boom, there were stories about landmen lining up at courthouses in oil patch counties before the doors opened, eager to trace the real estate transactions and buy up leases before their competitors got to the owners.[7] An article I'd read earlier in 2013 in the *New York Times* magazine described the scene in 2009 at the Mountrail County courthouse, not far from the original Iverson well. In cold weather, the landmen held their place in line with their briefcases while they waited in their cars, engines running, for a clerk to unlock the front door of the courthouse.

Burke County was just north of the excitement, on the edge of the oil fields, so the courthouse was never quite as busy in 2009. But I felt for any landmen dispatched on a cold morning to tiny Bowbells in the earliest days of the boom. Even a hot cup of coffee seemed scarce.

Back in their offices, the landmen and their support staff draw up lease paperwork to send to the people who own the mineral rights—if they can determine exactly who owns them. It's not this way in all states, but in North Dakota, the mineral rights are severable from the land. That means the people who own the mineral rights beneath the earth might be different from the people who own the surface rights.

It's known as a split estate. State laws favor the oil companies and the mineral holders, which can cause all sorts of obvious heartache.[8]

Sometimes, the people who live or farm on the land own no mineral rights and have no say in how or where or when oil companies decide to drill. They also get no share in the royalties from the oil being pumped out of the earth, even though the noise and pollution from fracking may disrupt their lives, their farms, or their health and tranquility. The United States is the only country in the world where private individuals—and not the government—own the rights to what is beneath the earth. Split estates are especially common in the American West, in part because of settlement patterns created by the Homestead Act. It's why whether to drill or not is often "an individual rather than a collective choice," wrote Colin Jerolmack in *Up to Heaven and Down to Hell*, a look at the effects of fracking in a Pennsylvania town.[9] But the consequences are almost always collective.

By 2013, many Americans were familiar with fracking and its potential consequences from *Gasland*, the 2010 HBO documentary by Josh Fox that chronicled the shale gas boom in Pennsylvania.[10] Among the film's most memorable scenes was one where a homeowner lit on fire the gas coming out of the tap in his kitchen faucet, underneath a sign reading "Do not drink this water." Fox, now a climate activist, began working on the film in 2008 when he got a $100,000 offer from a landman to lease nineteen acres he owned in northeastern Pennsylvania.[11]

Landmen developed a reputation as unscrupulous, one that their national professional organization sought to dispel by asking members to uphold a set of ethical guidelines. Most do.[12] But landmen are in the employ of the oil companies who want to drill on the land, not the people who own the land or the mineral rights. Fairly or not, landmen are notorious for persuading uneducated land or mineral owners to sign away their rights for cash up front, or to accept meager royalties. That day in the Burke County courthouse, the landman said he didn't want me to print his name or to say which oil companies he was working for. But he showed me how to find the paperwork I wanted to see, and he told me I could put any copies I made on his company account. I accepted his offer, happy for once to stick an energy company with my bill.

I began with the indexes, stored in a room with a fireproof metal door. Weighing at least five pounds each, the indexes were bound in cream-colored canvas ledgers about five inches tall and big enough to hold legal-sized paperwork. They were stacked fourteen books high, on steel shelves equipped with rollers that slid out for perusal and back in again for storage. Using the legal description of the township where Anna staked her claim, I found the book where a clerk recorded the original title to the land.

The real estate transaction records spelled out a dry history of the land and its settlement by white people, beginning with the original homestead patent granted to Anna by President William Taft. From there, I traced subsequent transactions, including leases. The first lease my grandparents signed in 1951 was with insurance agent Walter Braun and his partner, R. W. Porter, a duo who opened an office in Tioga within forty-eight hours of the first well on the Iverson farm.[13] Braun and Porter promised in the lease with my grandparents to drill no closer than two hundred feet to a dwelling, and to bury any pipes "below plow depth." If the company discovered oil or gas, my grandparents would have earned royalties amounting to one-eighth the value of what was coming out of the earth. As a landowner, you could also get natural gas from the well for your home or farm use if you supplied the pipeworks. This was a boilerplate lease clause on working farms in the 1950s.

The paperwork in the courthouse took me into the 1970s, when my grandfather sold the land that Anna once homesteaded. Ever since that day in 1951 when the Iverson well came in, my grandfather believed there was oil there. He couldn't let go of his dream. He wasn't sentimental about his childhood, but the land was the only physical connection remaining to Anna, the mother he never knew. So even though he sold the land in 1973 to a neighboring farmer for $18,500, Ed held onto a piece of his old dream. He kept 40 percent of any future mineral rights on the land, just in case.

This was how my mother and her brother, Ronnie, and two half-sisters they barely spoke to came in 2009 to own mineral rights in North Dakota. A landman with Empire Oil Company tracked them down on behalf of St. Croix Exploration Company and offered them a lease. It arrived in my mother's mailbox in 2009 just when she needed it most. By the time I arrived at the Burke County courthouse in 2013, Cornerstone Natural Resources owned the leases. It was a small, independent oil company based in Denver.

That first lease in 1951 paid my grandparents $160, roughly equivalent to $1,500 today. It was a tidy little windfall. My grandparents lived a relatively prosperous working-class life in Helena, but it was one still scarred by the scarcity of the Great Depression and the rationing of the war years. Among my mother's memories of growing up in the shadow of the Depression was the taste of powdered milk. She shuddered in revulsion once when she saw powdered milk in use at the home of one of my childhood friends. Later, in the car on the way home, she explained why. Powdered milk was one way of stretching a grocery budget and, to my mother, a lingering taste of the poverty and hard times her parents could never forget.

Even so, my mother's childhood was one full of postwar promise. Checks from the oil companies paid for the small luxuries of a 1950s girlhood: dolls and a pet poodle, 45s for her record player, accordion lessons. All were lovingly documented with a Kodak Brownie camera in black-and-white snapshots tucked into keepsake albums.

I saw those snapshots as proof that my mother was surrounded by love, and that her family wanted the best for her. Bill Shemorry's photo offered evidence of another kind, that there was financial freedom beneath the earth in North Dakota. My mother may not have known about Shemorry's iconic image, not directly. Yet what it represented was foundational to the family theology of windfalls. The possibility of striking it rich dangled in front of my mother throughout her childhood and teen years.

It was a pervasive doctrine in midcentury America, so widespread it was simply part of the atmosphere. National trade groups, backed by local boosters eager to lap up any extras, wanted to shape North Dakota into the next Texas or Oklahoma. The American Petroleum Institute even called North Dakota the "newest oil state."

In October 1951, just months after the discovery of oil in North Dakota, API took out a two-page advertising spread in *Life* magazine and *Collier's*. With a combined circulation of nearly nine million, plenty of Americans saw the financial promise of the headline: "Oil Rush in North Dakota." Subscribers may not have read the ad copy, which seemed mostly an effort to stave off federal regulation. But it didn't matter. The pictures and headline said it all: white men were drilling into the earth to build wealth. The message was explicit: you, too, might get rich if you invest in North Dakota.

The magazine spread included a posed photo of a landman "negotiating" with a North Dakota farmer. The landman is wearing pleated high-waisted khakis, a short-sleeved white shirt and a straw fedora with a dark band. He's clearly a city guy, plugging away at a portable typewriter set atop the chest-height wheel of the farmer's Allis-Chalmers tractor. The farmer, named as J. W. Tyler in the caption, has a straw cowboy hat pushed back off his forehead. Tyler's left arm drapes over the steering wheel of the tractor. He stares down the landman in a way that seems to say that he may look like a dumb country boy perched on his tractor, but he knows exactly what's up and he won't be swindled by the fedora-wearing city slicker clacking away at his typewriter.[14]

It may seem surprising, given all the hoopla around oil and the dreams of riches, but in some places in America, people were reluctant to turn over their wheat fields and bayous to the oil companies. Never fear! The oil companies were well prepared to make their case for a petroleum-fueled future. They deployed not only Madison Avenue's midcentury advertising might, but the tools of cinematic propaganda perfected by the American film industry during World War II.

Cinema had great power right before television took off in the 1950s, at a time when ninety million Americans went to the movies each week.[15] Among the most visually influential of the postwar oil boom films was *Louisiana Story*. Released in 1948, it's a scripted docudrama by Robert Flaherty, who is best known for the 1922 silent film *Nanook of the North*. Standard Oil paid Flaherty $258,000 to make *Louisiana Story*, which introduces viewers to a young Cajun boy and his family in the lush but impoverished swamplands of southern Louisiana.[16]

The black-and-white film depicts the extraordinary natural beauty of the Louisiana bayous. Flaherty portrays oil drilling as an engineering marvel, even as the film leaves viewers with a nostalgic but false impression that the simpler, nature-based ways of the bayou's past could coexist with the promise of oil riches in the future. *Louisiana Story* doesn't meet today's definition of a documentary. The corporate hand in the picture is too great, and it's a scripted production. But visually, it had great influence on how future documentary filmmakers tell stories about the plunder of natural resources. The film's story, written by Flaherty and his wife, Frances Flaherty, was nominated for an Oscar. The music was composed by Virgil Thomson and played by the Philadelphia Orchestra. The soundtrack won a Pulitzer Prize for distinguished musical composition, the only film score to ever do so.

By 1953, oil companies understood they could sway public opinion with film. Produced by the American Petroleum Institute, *American Frontier* was a scripted black-and-white documentary about the North Dakota oil boom, directed by Willard Van Dyke, an influential filmmaker of the era who went on to lead the film department at the Museum of Modern Art in New York. Like *Louisiana Story*, the film is a beautiful depiction of a place—only this time it's the winter-swept prairies of western North Dakota. And like *Louisiana Story*, it's a big-budget endeavor backed by oil money; its score was performed by the New York Philharmonic Orchestra.[17]

Among my favorite scenes in *American Frontier* is one where Barbara, the main character's wife, serves coffee to some landmen who visit the family farm near Williston. Barbara is not seated at the table; she's relegated to making coffee. But it is Barbara who we see smile when the landmen offer ten cents an acre to lease the land for ten years. Coffee pot in hand, she asks the smartest question of anyone at the table, including her husband, Nils: "And, if you should hit oil?"

You'll get the standard royalty, the landman tells her: one-eighth of every barrel. Her father-in-law remains skeptical. "Wheat is good enough for me," he says, in a thick Norwegian accent. The scene borrows heavily from *Louisiana Story*, which also features a kitchen table coffee visit with a landman. The kitchen scenes in both films feature folksy jokes from the crusty white patriarch with a foreign accent, a man who settled the so-called frontier, and who will eventually sign the lease despite his reservations about what it means for the land he values most. Because who in America in the 1950s says no to an oil well in their back yard?

It's possible the films played at the Bowbells movie theater or one of the other cinemas in nearby towns. Many of the tiny towns in the area boasted movie theaters, built in the 1920s with the same optimism of the Burke County courthouse. In fact, many tiny towns in rural North Dakota have single-screen cinemas in operation to this day on their main streets. The Bowbells movie theater burned down in the late 1950s, but there was still a theater in Kenmare, just sixteen miles away. I'd driven through Kenmare a few days earlier and had been so struck by the theater's existence that I took a photo of it. *White House Down* was playing, starring Channing Tatum, Jamie Foxx, and Maggie Gyllenhaal.

My mother must have seen *American Frontier*. She was eight years old when it came out, and a formative experience of her childhood was to spend Saturdays at the movies with her brother. Cartoons, newsreels, westerns, sci-fi flicks...she saw them all. Films like *American Frontier* would have been on the bill, especially in oil patch states. These sorts of industrial films were shown widely in their heyday. People saw them in schools and at Kiwanis Club meetings and in union halls and before feature films. They were often very high quality, with imagery and persuasive storytelling that lingered long after the credits ended. And westerns were so popular in the 1950s that people might have purchased a ticket to *American Frontier* thinking they were about to get a cowboy yarn with their popcorn.

But when the oil companies pay for the picture, it's no longer merely entertainment, it's persuasion. It's industry propaganda, appropriating powerful visuals to shape the midcentury prosperity doctrine that seemed to seep into all aspects of postwar life in America.

The stories in these films contributed to the myths around the collective American yearning for wealth and prosperity. The oil boom of the 1950s merely appropriated long-standing myths of westward expansion, the same stories of American exceptionalism that accompanied my great-grandmother Anna when she ventured to North Dakota in the early 1900s. After all that prairie had been broken, the search for oil was merely the next, new frontier.

For my mother, the possibilities promised by that first North Dakota oil boom in 1951 were likely a point of pride. She wasn't poor, but she came from a working-class background. As a teenager, she faced the daily mortification of being the daughter of the custodian at her junior high school. Even though everyone thought of her father as the guy in coveralls who maintained the boiler and mopped the school floors, my mother got to walk around with secret knowledge: someday she could be an oil heiress. It's very human—and very American—to dream of the ease we think comes with riches, whether you're Barbara with your coffee pot, or a kid in Montana in the 1950s who watched a lot of movies on Saturdays.

————

The Burke County courthouse closed at 4 p.m. I exited with my paperwork, craving a hot cup of coffee for my two-hour drive. Bowbells had one bar as far as I could tell, the Black Gold Saloon, and it wasn't open. A café on the outskirts of town posted a sign saying it was closed for the season. I couldn't tell whether August was out of season or whether the restaurant had been closed so long that multiple seasons had passed by with the same sign on the door. There was no grocery store, either. I regretted not buying a slice of pizza I'd seen under a heat lamp at the Cenex station fifteen miles away on Highway 52, where I stopped to use the bathroom and to fill up my gas tank. It was one of the few places nearby that seemed to offer anything approximating a hot meal.

As I drove back toward Minot for the night, I thought about what it was like in North Dakota in 1951 as oil mania swept the state. It must have been seen as an extraordinary opportunity. There wasn't much of an environmental movement back then. The sole naysayers in the 1950s were farmers worried about the effects on their fields, and even then, they leaned toward drilling for the riches beneath the earth.

My grandfather was no different, looking on from Montana. He wanted to reap from the leases what he could, for his family. Maybe I was rationalizing my family's past decisions, but I took some satisfaction in knowing that my mother's education was indirectly tied to a beer-drinking newspaperman who grabbed his waders and drove through floods for his photo. Because people saw Bill Shemorry's photo in 1951, investors took the North Dakota boom seriously. Because my mother had money from the oil leases to pay for college, it was much more likely my sister and I would go to college, too, and that I would go on to be a journalist. The money from the oil leases mattered. It shaped our lives. It was shaping my life even as I drove the backroads of Burke County. The time in the car gave me space to think. Already, I knew that when I got home, I was going to quit my stagnant job.

As a journalist, I understood what Shemorry was after that night in 1951 when the Iverson well came in. He was in search of his own windfall, even if it was indirectly connected to the industry. An oil boom was hopeful, especially after the hard years of the Depression and World War II. It was exciting news to cover as a small-town journalist. It meant a newspaper boom. Oil companies would pay to run want ads in the classified section of the newspaper. New restaurants and car dealerships and department stores and movie theaters would open, and their owners would want to buy ads, too, to reach all those new people who were buying things to build the boom. More people also meant more potential subscribers. It would mean general prosperity, community wide.

Looking back, we see with modern eyes the climate catastrophe foretold in the flare that illuminated Shemorry's photo of the first oil well. The satellite photos of the Bakken boom of the 2010s illustrated an even more calamitous future. They both were powerful images that shaped perception, just like the oil company–sponsored movies did. Even for those who don't know that it's called a pumpjack, the sight of one bobbing in and out of the earth immediately calls oil fields to mind. The emotion that accompanies the sight may vary, depending on your point of view.

Shemorry knew exactly what the sight of that derrick meant in 1951 to people like my grandfather who had been very poor just a decade earlier. In the 1950s, oil derricks and pumpjacks still looked and felt like money, not the destruction of the planet.

5
PAST IS PROLOGUE

September 2013

North Dakota crude: *$96.69 per barrel*

ANNA AT TWENTY-SEVEN WAS PLUMP-FACED AND PRETTY, WITH SEARING eyes that re-appeared generations later in the blue gaze of her descendants. On a cold but clear mid-December day in 1905, the Reverend Nicolay Nilsen of the Lutheran Free Church married her to Andrew Haraseth.

Her new husband, nineteen years her senior, was seated in their wedding photograph, his right elbow resting on an armchair. Anna stood next to him, the better to capture for future generations the simple splendor of her white wedding gown. The high-necked dress had long, puffy sleeves, and in the fashion of the time, it bloused out from the nipped-in waist. Three neat pleats on each side of the skirt fell to a precise, floor-grazing hem. Anna's hair was piled on top of her head in a Gibson Girl pompadour, and she wore a spray of white flowers on each side of her center part. She pinned a corsage of darker flowers to the upper left bodice of her dress, near her shoulder. Andrew wore a dark three-piece suit and a sprig of pale flowers on his left lapel.

In the photo he's a trim man with a full head of thin but neatly combed hair, and a bushy handlebar mustache.

Anna and Andrew Haraseth on their wedding day in Minot, North Dakota. December 13, 1905.

Anna and Andrew walked out of the photography studio in their wedding outfits and into the height of the final big land boom in North Dakota's homesteading era. Their photograph suggests the sort of prosperity that allowed them to have a portrait made wearing elegant clothing, and the money for hothouse flowers on their wedding day. But it told me little else.

My own marriage was only six months old on the day I headed to the National Archives in Washington, DC, to search for Anna's homestead paperwork. Like Anna and Andrew, Chris and I had also been married on a cold but clear day. Our wedding was in late March. More than a hundred of our friends and family gathered around us in a circle in the World War memorial on the National Mall in Washington, DC as we exchanged vows. I, too, wore a carefully hemmed gown. The dress dragged in the dirt during the ceremony. By the end of the night, the bodice was smeared with chocolate cake. Anna, I suspect, was far more careful with her gown.

I can't know whether my great-grandmother and her husband were happy or whether they were in love—everyone looked so serious in photos back then. And yet, whenever I looked at their wedding photo, I couldn't help but imagine them basking in at least a small measure of the joy of the early months of my own marriage.

All I really knew about Anna, though, was what I'd gleaned from a public records search and from old newspapers. She was born in Elizabeth, Minnesota, in 1878. She was confirmed at a Lutheran church in 1893 when she was fifteen years old. She married Andrew in Minot in late 1905. She and Andrew had a baby almost exactly a year later—my grandfather, Ed. A perfunctory search of the federal Bureau of Land Management's homestead database showed that President William Taft granted her a patent for her homestead in 1912. I'd seen a duplicate of that document in the courthouse in Bowbells. I knew Anna died in 1921, but I had no idea why or where. I was still waiting on a clerk in North Dakota to send me a death certificate.

Anna left few other traces. No diaries and no letters. Just the wedding photo and two pieces of jewelry: a gold brooch inlaid with five pearls and a gold-plated pocket watch. And, of course, the mineral rights and my mother's unreliable accounts of Anna's life as a homesteader.

Family lore was this: Anna homesteaded her land by herself, as a single woman in her late twenties, before she married Andrew. Anna, so the lore went, was a homesteader who overcame great adversity to establish a home and farm in a harsh climate in exchange for the opportunity promised by land ownership.

Like other homesteaders, Anna got 160 acres of land from the federal government for a $14 filing fee and the promise to live on the land, build a structure, and cultivate it for five years. When settlers made good on their claim, they had to file a personal narrative explaining what they did to improve their land. They had to explain what buildings they erected, what fields they cleared, and what crops they sowed.[1]

All of those documents are stored at the National Archives building on Constitution Avenue between the U.S. Capitol and the White House in Washington, DC. Completed in 1937, the grand marble building with Corinthian columns sits in a flood plain of the Potomac River, on a site that climate scientists predict will one day be inundated with water as sea levels rise.[2]

For now, though, tourists line up to enter beneath a pair of imposing thirty-eight-foot tall bronze doors guarding the original Constitution, the Bill of Rights, and the Declaration of Independence. Underbuilt from the start, it took just thirty years for archivists to fill it to capacity. That means much of the nation's history is now stored in a glass-and-concrete building with 520 miles worth of shelving in suburban Maryland. Only 2 to 3 percent of federal records are retained each year, and someone long ago decided that the ten million homesteading case files were worthy of keeping—in the fancy downtown temple, no less, not a satellite archive.[3]

Over the years, I'd walked by the National Archives hundreds of times without paying it much attention. The building sits directly across from the sculpture gardens at the National Gallery of Art, a place I had visited with my friends for jazz concerts on plenty of Friday nights in the summer. Until this visit, I'd ventured inside the Archives just once, as a tourist on a sticky July day to see with my dad the three American documents historians call the "Charters of Freedom."

This time, though, I entered through the less formal but no less ornate back entry on Pennsylvania Avenue. Two statues, rising twenty-five feet above the sidewalk, flank the rear entrance. I paused at *Future*, sculpted on site by Robert Ingersoll Aitken from Indiana limestone brought on flatbed railcars to the capital. Aitken's modern muse sat with a blank book in her lap: the future as yet unwritten. I got chills reading the inscription from Shakespeare's *The Tempest*, carved into the base: "What is past is prologue."

It seemed so prophetic. I had just quit my newspaper job to pursue Anna's story. My visit to the archives was my first official day of freedom from a career I loved but a job I'd come to loathe. I finally got to go in search of the story I'd long wanted to tell. I had a blank book to write the story in, using all the skills and experience I had cultivated as a journalist. It was a brave new world. I was as optimistic as I'd ever be, blissfully unaware of what "Future" planned to write in my book.

Inside, I passed through a metal detector and beyond the guards, who were there to search outgoing visitors for pilfered documents. I watched a short video orientation and applied for a research card, which allowed me access to the main research room. Cameras, scanners, and laptops were allowed, but no pens, only pencils. I stashed my backpack in a locker and headed to the walnut-paneled research room.

Once inside, I requested the tract book used to record homestead transactions by the General Land Office, the precursor to the federal Bureau of Land Management. From it, I could find the coordinates on Anna's homestead to request the specific case file connected to the transaction. I wrote the file numbers down on a pull slip. I searched not just for Anna, but for her husband and their neighbors. I was hopeful that Anna's file would include a description of her life on the farm, either in her own words or those of the people who testified as witnesses for her claim. Mostly, I was in search of anything that would shed light on what Anna's life was like.

Archivists retrieve records requests from the stacks just five times daily, so I waited quietly at a table for my files to be delivered on a cart. The tall-ceilinged room was soft with natural light from windows that took up most of the street-facing wall. People bowed over their documents, illuminated by pools of light from the milk glass shades of the brass library lamps. The room smelled like one of those paperboard boxes that once held negatives or photography slides. In this hushed temple to history, people knew of card catalogs, the Dewey decimal system, mimeographs, and microfilm. Nothing, not even an unexpected find on Google, could replace the anticipation of unearthing a tactile record from the past.

While I waited, I read through a pamphlet on the land records stored at the Archives. The pamphlet suggested that when "used imaginatively," even the smallest homestead case file offered a treasure trove of research possibility. "Depending upon the type and time period of land entry, the case file may yield only a few facts already known to the researcher, or it may present new insights about ancestors, family history, title, and land use issues."[4]

President Abraham Lincoln signed the Homestead Act into law in 1862 in the precarious early years of the Civil War. By the time Anna claimed a homestead, the law was well into its second generation.

American leaders had long sought a way to settle vast swaths of the North American continent acquired via conquest, purchase, war, and other means. Congress had been considering versions of the Homestead Act since 1850, but as the Civil War began, it became even more imperative for Lincoln to pass the legislation. Even as he sought to expand access to economic opportunity, Lincoln wanted to keep slavery (and sympathetic southerners who supported slavery) out of western states. One way to do just that was to encourage Union veterans of the Civil War to settle on newly available homesteads. And so, with southern senators no longer in the way after the South's secession, Congress readily passed the Homestead Act.[5]

Lincoln also signed into law three other bills aimed at creating institutions, policies, and even transportation and education systems that favored the regions open to homesteading and that made it harder for slavery-friendly southerners to settle the settle the West. Days before the Homestead Act became law, Lincoln signed legislation that created the Department of Agriculture. A short time later, he signed the Pacific Railway Act, which authorized a transcontinental railroad that traced a northern route across the continent, instead of through the South. Next came the Morrill Act, which gave states title to western lands they could sell to pay for the construction of universities, so-called land-grant colleges. Lincoln also signed the Emancipation Proclamation, making it clear that the war was about abolition of slavery.[6]

The Homestead Act changed in its particulars over the years, but at its core, it represented a wealth-building opportunity for people of little means who otherwise were without access to land or capital. "The beauty of the Homestead Act was that anyone willing to move west and stake a claim was eligible for the public land," wrote Trina Williams Shanks, a scholar who has studied the asset-building potential of the Homestead Act as well as how modern social programs could replicate its successes. "Wage earners, recent immigrants, young adults from large families desiring private property, those seeking adventure, those wanting to shape politics or culture in the 'lawless' West, those seeking a new start, and thousands of others could all pursue their personal dreams and aspirations."[7]

In theory, most Americans could access the Homestead Act. Married women couldn't claim a homestead, but single women and widows could, and many did, especially in Anna's time in the early 1900s. The government-sanctioned ability to stake a claim was a tremendous economic and social freedom for women during and after the Civil War when they remained sixty years away from even being eligible to vote in national elections. Immigrants in the process of naturalization who met all the other requirements could also claim homesteads. And of course, single and married men could claim homesteads; husbands were considered not just heads of the household, but responsible for their wives.

Black Americans could also claim homesteads, although the process was rarely as straightforward as it was for white homesteaders.[8] Racist laws and exclusionary practices in many western places made much of the region hostile to any arrival who wasn't white. Sometimes the laws in "sundown towns" were explicit, in the form of notices posted at the town limits warning Black people to leave before dark. Other times, they were more informal. Despite long odds, researchers estimate that about thirty-five hundred Black homesteaders filed successful claims on as much as six-hundred-fifty thousand acres of land.[9]

Native Americans could not claim land under the Homestead Act; North America's first people were not legally considered citizens of the United States until 1924.[10] Before Anna and Andrew's marriage and arrival in northwestern North Dakota, the land was inhabited mostly by the Assiniboine people. The prairies were not empty; they had never been empty of people or wildlife. But when Anna and Andrew arrived in northwestern North Dakota, most Assiniboine were living on the Fort Peck reservation in northeastern Montana. The Assiniboine population was diminished by disease and warfare as well as the effects of broken treaties and other less formal but equally terrorizing tactics committed by and in the name of the U.S. government. Those tactics included starvation. Samuel E. Snider, one of a series of hapless federal agents in charge of the Fort Peck reservation in the 1880s, wrote that "with no crop, no game, and as yet no supplies, the wolf of hunger is in every lodge." Three hundred Assiniboine starved to death in the winter of 1883–1884 as bison were hunted to near extinction, according to a chapter of a tribal history entitled "The Starving Years."[11]

It is likely that by the early 1900s, many white settlers in northwestern North Dakota, including my great-grandparents, believed it was their God-given, government-approved right to claim land no longer occupied by the Assiniboine or other Indigenous inhabitants. It is unlikely they considered why it was no longer inhabited.

Over the 123-year existence of the Homestead Act, four million homesteaders settled land on two-hundred-seventy million acres in thirty states.[12] Although the land available for homesteading was nearly free for those who could access it, there were many barriers to a successful homestead. Beyond the $14 filing fee, people needed money to travel to the newly available lands. They needed to pay to feed and clothe themselves or their families until their farms began paying off their early investments. They needed money to build a structure on the land, as required by law. They needed cash or credit to buy livestock and the seed necessary for a first season of crops.

Those who were successful were often surrounded by a community that could share tools, equipment, or labor—whether it was fellow immigrants, extended family, or those of a like-minded religious persuasion. And even then, the homesteaders of the Great Plains faced grasshoppers, drought, flood, blizzards, range fires, and the same other natural hazards and hardships that modern farmers continue to confront.

Anna's time as a homesteader, I knew, was not quite as difficult as those who made claims in the years immediately following the Civil

War. Telephones, electricity, and automobiles existed as did an extensive railroad network. Nonetheless, the land Anna claimed in northwest North Dakota was not well settled or ever that populated. It was
far from established towns, for one, and it drew few white settlers until
the Great Northern and Soo Line railroads began competing with each
other in the early 1900s to build branch lines into northwestern North
Dakota and northeastern Montana.[13]

The railroads were eager to exploit new farming techniques to help
farmers improve their yields. More grain meant more traffic along
their rail lines and therefore more profit. The Great Northern in 1906
even ran a train in the Dakotas and Minnesota called the Good Seed
Special featuring farming experts from regional agriculture colleges.[14]
Dozens of new towns popped up in northwest North Dakota in this
period, including the small town of Larson near Anna's homestead.
Many of the towns were marked by new grain elevators, the skyscrap-
ers of the prairies.

The railroads encouraged settlement with aggressive advertising
campaigns touting the potential riches to be made growing wheat and
other grains on the high, northern Great Plains. Boosters exaggerated favorable growing conditions that, it turned out, would not last.
Cyclical drought was a feature of the prairie, not an anomaly.

Newspapers of Anna's time abounded with success stories of the
white settlers who sought out the prairies of North Dakota in the final
land boom era. I was particularly drawn to the stories in the *Ward
County Independent*, an unabashedly boosterish newspaper that sali-
vated for growth in all its forms. The newspaper loved to downplay the
cold climate. "An eastern person has only to look through the North
Dakota newspapers and read the accounts of the number of buildings
being erected this winter, to be satisfied as to the climate of our state,"
the paper wrote in February 1906. "When carpenters can work in the

middle of the winter without gloves or mittens, certainly [*sic*] can not be a very cold country."[15] On the editorial page of its final edition of 1905, the *Independent* offered up a wish for prosperity in 1906: "A New Year's Resolution: Resolved that I will boost North Dakota in every turn of the road during the coming year."[16]

The publishers of the *Independent* bragged about the county's population ballooning from 7,961 to 33,468 people between 1900 and 1905.[17] It was mind-boggling growth, beyond even the modern-day boom experienced in places like Williston. The newspapers flourished in part because of all the legal advertising required by federal law whenever someone proved their homestead claim after five years. Fueled by the railroads, a land boom was also a newspaper boom. "The booms created the newspapers, and the newspapers, in turn, did all they could to help the booms," wrote Elwyn B. Robinson in his history of North Dakota.[18]

The *Independent* had no mention of Anna and Andrew's wedding in the section of the newspaper where editors printed the names and ages of local couples who applied for a marriage license. But there was a news item about the then-president's daughter, Alice Roosevelt, and her fiancé, the Republican congressman Nicholas Longworth of Ohio. The two-decade difference between Roosevelt and Longworth raised eyebrows, but it wasn't especially unusual. "Alice is 21 years of age while her intended is nearly forty, but he is a good sensible fellow and worth a hundred titled foreigners. Alice has shown herself to be a good sensible American girl."[19]

A young woman who didn't prove herself to be a "good, sensible American girl" would have soured quickly into pitiful old maid in Anna's time. And yet, Anna and Andrew would both have been considered old for a first marriage. A woman in Anna's time who married at twenty-seven gave up a decade's worth of wide-ranging adult independence that was unavailable to married women. Marriage in 1905 might have come with the financial protection of a partner and the then-crucial respectability of the title Mrs. in front of a woman's name. But it was no guarantee of a good life, particularly at a time when women had few family planning measures.

The beauty of modern marriage, it seemed to me, was that you willingly decided to share your life and finances and hopes and dreams with someone else, on equal footing.

At our wedding, Chris and I walked down parallel paths to the memorial, embracing the symbolism of coming together to walk up the steps as a pair. Chris hugged his mother, and I hugged my father, but my father didn't give me away. At thirty-nine, I'd moved beyond young woman. But I also rejected the language of soured old maid. Yes, I wore a white wedding gown. I never considered changing my name, the one I'd been given at birth and the one that appeared atop thousands of newspaper articles. Chris never brought it up—he knew how I felt. My independence was hard earned by the women who came before me, and I didn't take it lightly.

A popular subgenre of the homesteading success stories I read in the *Independent* and other newspapers of the time were those of the spirited young white woman who made good on the prairies. I was especially intrigued by a story I read about a woman named Clara Kress, from 1905. "Miss Clara Kress is a very good example of a plucky North Dakota young woman and by her grit gained a valuable 160-acre farm near Deering, which is easily worth $2,500," the *Independent* wrote. "During the entire summer, fall and winter she has remained at her post and just recently made a solid proof of the claim."[20]

From my mother, I grew up understanding I was descended from a woman not unlike Kress, someone who, on her own, settled the untamed prairies of North Dakota at a time when an independent woman of determination was newsworthy in a local paper keen on boosting community fortunes.

But as I read more of the newspapers from the era, a darker story line also emerged. It was about women who went shack wacky in the winter in small cabins on the prairie. The same newspaper that reported on the pluck and grit of young women like Kress also published accounts about people, mostly women with young children, who lost their minds on the prairie.

The papers often published stories about those who'd gone to Jamestown, a euphemism for being accompanied on the railroad by the county sheriff to the state asylum, perhaps in a straightjacket: "Mrs. Mattie Sjaastad, who lived ten miles south of Stanley, was taken to the Jamestown asylum yesterday by H. D. Shepard," the *Independent* reported in 1907. "The poor woman has not been well for months and at last her mind became effected [*sic*]. She has a ten month's old [*sic*] baby. The case is pitiful."[21]

Other women who homesteaded on the North Dakota prairie during this time kept diaries and letters, and wrote about being able to see visitors coming on the horizon. Life was hard but "homesteaders knew how to have fun," wrote H. Elaine Lindgren, in a history she compiled of women homesteaders in North Dakota.[22] I read about these white women, and they always had time to change their dresses, to comb their hair and to "get a good meal in running order before they arrive," as homesteader Bess Cobb wrote to friends in 1906 about life on her homestead in North Dakota.[23] The shack parties sounded like an especially good time. Effie Vivian Smith lived in the same county as Anna, and in 1906 wrote of seven people showing up at her 10x16 cabin for a party one winter night. "I never enjoyed myself better in my life than I have this winter," she told a cousin. "We go some place or some one is here from 1 to 4 times a week."[24] And Anna Zimmerman wrote of riding twenty-five miles on horseback to dances, dancing all night, and then riding home. "With the shortage of girls, I often played the harmonica and danced at the same time," she remembered.[25]

Based on my mother's stories and my own research, I began creating a version of Anna. In my mind, she was a woman of daring and grit who embodied the myths my mother and my culture fed me. I saw her through the rosy lenses of not just a newlywed but through the eyes of a modern-day woman who'd been fed stories since childhood about the adventurous spirit of the American women who claimed homesteads.

There at the archives as I waited for my files, I imagined Anna as a generous hostess who donned a pretty calico dress when visitors showed up at her shack. I envisioned her as game for a fifty-mile round-trip ride on horseback, just to dance, like Zimmerman. I wanted an ancestor with an exceptional story, one who defied the limitations society put on the women of her time. I imagined Anna as a youthful and hopeful young woman, in a world made by and for men.

When the case files arrived, I unfolded the legal-sized records with caution, eager for evidence of my imaginary prairie goddess. The documents appeared to have been unopened since they were sent from the General Land Office in Minot to Washington, DC, a century earlier. The last person to handle the papers assembled them neatly, in packets.[26]

There in the quiet archives, I paged through the unfolded packets carefully but quickly, the journalist in me looking for the breaking news first. I gasped out loud when I read it on the page:

"She is insane."

6

MANIFEST DESTINY

September 2013

North Dakota crude: *$96.69 per barrel*

IT SEEMED I HAD INHERITED DUELING NARRATIVES OF WOMANHOOD: INDE-pendence served up with a side of insanity. Plucky *and* pitiful.

I sat at the walnut table in the research room at the National Archives, piecing together Anna's life in her early twenties. An unexpected story began to emerge from the affidavits and legal correspondence in her file. The account I read was one that challenged everything I thought I knew about Anna and her time as a homesteader—and my family's history.[1]

In 1905, Anna was working as a dressmaker in Fargo. She lived with her older sister Christine, who was married to a postal agent for the railroad, George Langen. Anna's sister and brother-in-law had three children of their own. Anna was saving money for a homestead, with the intention to "make it her future home and create her independence," her lawyer Henry Krogh wrote in an affidavit she filed with the General Land Office.

Fargo was a rapidly growing city of more than ten thousand

people, still in the process of rebuilding after a fire burned the city's core nearly to the ground in 1893. There were new streetcar lines, telephone exchanges, and elegant new hotels and restaurants that served delicacies brought in daily on the Northern Pacific Railroad from larger cities.[2]

And yet Fargo was also a dry town. All of North Dakota had been dry since statehood in 1889, when men voted narrowly to outlaw the sale of booze in the state constitution. The Women's Christian Temperance Union was well-organized and influential in North Dakota, and if women had been able to vote in 1889, statewide prohibition likely would have passed with an even a larger margin. Nonetheless, illegal saloons known as blind pigs were popular and ubiquitous throughout North Dakota until prohibition ended nationally in 1933.[3]

But just across the Red River in Minnesota sat Fargo's sister city, Moorhead, and it was packed with saloons and plenty of visitors from North Dakota. Moorhead was home to only thirty-seven hundred people in 1900, but it had forty-seven saloons and a brewery until it, too, went dry in 1915. It wouldn't have been respectable for Anna and other single women like her to take part in the revelry, but they must have been familiar with Moorhead's jag wagons, horse-drawn taxis operated by the saloons to ferry thirsty men over the bridge to Minnesota and back.[4]

Anna and her friends and family were far more likely to have attended the wholesome entertainment associated with their church, like baseball games and ice cream socials. Anna's sister Christine regularly hosted meetings of the Ladies Aid Society of the English Lutheran church at her home, according to announcements I found in old newspapers.[5] I suspected the whole family would have attended a Fourth of July picnic put on by one of the Lutheran churches in town in 1904. The picnic raised one hundred dollars, which was enough to get a write-up in the *Forum*. "An elegant dinner was served to a large crowd of people," the paper reported. "In the afternoon a fine speech was delivered by Professor Haugen of Decorah, Iowa. After the speech the crowd went to the ball game and then came back and ate ice cream and cake."[6]

But Anna also was likely drawn to an evening with one of the many itinerant preachers, teachers, entertainers, and mystics who made a living in the railroad towns of the Great Plains. By 1904, when Anna lived in Fargo as a single woman, occult studies were mainstream entertainment. Yoga, or at least yogic philosophy, had long been a part of the circuit, beginning in 1893 when Swami Vivekananda traveled to Chicago from India to speak at the World's Parliament of Religions. Vivekananda was the first introduction many Americans had to Hinduism, and he wowed audiences with his message of religious tolerance. He ushered in a wave of imitators, some of whom, like him, were legitimate spiritual guides. Others were complete quacks.[7]

That winter, a fortune-teller known as the Great Hindoo Seer turned up in Fargo on his way to represent India at the World's Fair in St. Louis. Notices of the impending arrival of the man they called the Hindoo Prince began showing up daily in the *Forum* in late February. The prince, who the newspaper boasted was "everywhere regarded as a master for his mystic art," arrived to great fanfare. He rented a suite at the Webster Hotel and he even installed phone lines so people could make appointments with him. Palm readings cost one dollar, a princely sum the equivalent of twenty-five dollars today. "Love, Not Gold, Sways the Divine Powers of the Great Hindoo," one headline read. (But pay your dollar, please!)[8]

The rooms at the Webster Hotel were for weeks crowded with fortune seekers. The Hindoo Prince even told fortunes on Sundays so that working people could call upon him. Anna must have. She lived just a few blocks from the Webster Hotel. I could imagine her waiting with the other seamstresses and clerks of Fargo in a parlor at the hotel for a palm reading on a Sunday. She would have worn one of her neatly sewn dresses, her precise clothing an exterior marker of her class aspirations.

"There are a good many people, especially ladies, in Fargo today who are awaiting the result of the prognostications made by the prince, which he declares will come to light before he leaves the city," the newspaper predicted.[9]

Like any single young person, if Anna got her fortune told, she must have asked about her love life. With his "brilliant eye" that "seemed to pierce to the very depths of all things," the Great Hindoo Seer would have looked Anna directly in the eye, boldly, in a way in which she was unaccustomed.[10] I imagined his fortune for her: "*You will face tragedy, but you will leave a legacy of black gold.*" It would have been cryptic, yet true.

There were only so many palms to read in Fargo, and the Great Hindoo Seer disappeared from the pages of the newspaper by mid-March of 1904. He no doubt made his way to other towns along the railroad, making his fortune one dollar at a time on his way to St. Louis, if that was in fact where he was going. I doubted that the Great Hindoo Seer was a prince—or Indian. He was most likely a charismatic confidence man who sold the assurance all humans seek, albeit cloaked in Orientalist garb fashionable in early 1900s America. He wasn't swindling anyone, either. Americans are the "most simple people in the world to read," as the paper noted, because we are in on the swindle. We know the confidence man is mostly selling us confidence, and we're willing to pay for a vial of it. Especially those of us who will do anything to manifest our dreams. We're nakedly greedy for news of the good fortune we've been taught to believe is our birthright, if only we work hard enough.

For most people, fortune-tellers do little more than confirm existing intuition about love affairs, business dealings, and yes, even political inclinations. But others might have interpreted a fortune-telling session as permission to push forward with their dreams, no matter who or what stood in the way. If you believed the prosperity gospel and the American exceptionalism of the times—and why wouldn't you?—there were fortunes out there for the taking. *Get rich now*, the fortune-tellers seemed to say, before someone else gets to the claim first. Or before it all dries up and blows away.

It was long that way in Dakota Territory, a place where gold rush gamblers stampeded the shimmering hills, where settlers plowed over the rich land with manifest destiny. "Manifestation" in America was often accompanied by cruelty, whether it was directed at the Indigenous people who made their home on the Plains before the arrival of white settlers, or extraction from the very land and its resources.

Fargo of the early 1900s was a town full of land schemers, their eyes firmly fixed on the real estate opportunities in homesteads on the western horizon. "A mania to secure land seized people," wrote Elwyn B. Robinson of 1905.[11] Anna was, no doubt, not the only young person in Fargo saving up for a homestead in the early 1900s. The *Fargo Forum* was packed with stories about homesteading, including speculation about whether Congress would authorize large-scale irrigation projects in parts of the arid West to make more farmable land available to settlers. The classified ad page of the *Fargo Forum* overflowed with real estate listings for agents who could help people find homesteading opportunities. Even Anna's lawyer Henry Krogh regularly advertised land deals in the newspaper. For $125 to $250, he offered "choice homestead locations—if taken at once."[12] Despite Krogh's sense of urgency, a great deal of land remained available in western North Dakota, the *Forum* noted. At the beginning of 1904, there were 8.5 million acres of federal land available to homestead in the central part of the state alone, the paper reported, underneath this promising headline: "Homes Open for Settlers."[13]

The land frenzy was palpable. One man in 1905 placed a classified ad in the *Fargo Forum* seeking to sell a home decor shop in the town of Mohall so he could have the money to seek his fortune on a homestead claim. It was the only such business in the town, he said in his ad, and there was plenty of work hanging wallpaper and painting homes. The reason for selling? "Owner must leave to live on his claim."[14] The federal government was strict about homestead claims. People were required to live on their land for most of the year for five years, and then in later iterations of the law, just three. The government required homesteaders to show they had made improvements, including cultivating the land and building a residence.

By 1905, Anna was itching to find a homestead of her own. She filed a homestead claim, but it was not the one I was familiar with, not the one that led to the mineral rights my mother inherited. The claim Anna filed in February 1905 was in east-central North Dakota, near the community of Medina. It was much closer to where Anna lived in Fargo.

In April 1905, Anna went out to look at her homestead in Medina. Her plan, her lawyer wrote, was to "make arrangements for building a house and establishing residence." But when she arrived, it was not the land she was first shown in February by people she described in an affidavit as "interested parties whom she believed and had had every reason to believe were honest and competent."

Neighbors pointed out the true cornerstones marking the boundaries of her property. The land she purchased was "composed of sand and stones practically entirely, with no grass on it to speak of, and entirely unfit either for farming or grazing purposes."

She "left the land in disgust," her lawyer wrote. A subsequent visit to Medina in June showed that the land was "worthless, and she could not possibly make a living on it in any way." She is "an unmarried woman 26 years old and dependent entirely upon herself and exertion," her lawyer wrote. Anna took the land in "good faith," for the "sole purpose of settling thereon and improving the same and making a home for herself."

All of this was in an affidavit Anna filed in October of 1905 with the Interior Department. She was asking for permission to give up her initial claim near Medina and to instead pick out another.

This was an audacious request. People generally got only one crack at a homestead. In addition to her filing fee, "she has been to great other expense in connection with said filing," her affidavit said. But "worst of all, she has lost her homestead right of which she as an American citizen felt so proud and had so much confidence in."

Perhaps the line was boilerplate her lawyer inserted in every appeal he filed with the General Land Office. Yet I wondered if it was accurate, or if Anna's family was trying to get her out of a rash commitment to the wrong land. Anna's affidavit was signed by her brother-in-law and one of his neighbors. The two men did not accompany Anna to the homestead in Medina, but they did attest to the truthfulness of her financial circumstances and her account of visiting the land.

It took two lawyers and several months of correspondence with the assistant commissioner at the General Land Office at the Interior Department in Washington, DC. Finally, though, Anna got approval to pick out a new parcel on November 2, 1905. She was still a single woman, still eligible for her own homestead.

And then, on that cold day in December 1905, Anna Josephine Sletvold donned her perfectly hemmed wedding dress. With marriage, she became Mrs. Andrew Haraseth. It was a decision that made her thereafter ineligible for a homestead in her own name.

The family lore was not ours to claim. Anna didn't homestead on her own on the prairie. Within just a few weeks of claiming her homestead, she got married.

Her husband, Andrew, was a Norwegian immigrant who settled in the small town of Elizabeth, Minnesota after his arrival in the United States. Many people from their small, western Minnesota town moved west on the railroad to North Dakota at the same time as Anna. It turns out that many of those people settled in Burke County, including Anna's father, one of her brothers, and a sister. Anna was not alone. She was surrounded by a community not unlike the rural one where she was raised in western Minnesota.

The Haraseths arrived on their homestead on July 10, 1906. By then, Anna was in the early stages of pregnancy. This is not in the file, but the math was obvious. She and Andrew got to work. They broke seventy-seven acres of virgin prairie that year. There is nothing in the file that describes the back-breaking labor of their early days on the prairie. But I imagine they were hopeful. They had the beginnings of a successful farm. It was summer, and the days were long. They had a baby on the way, my grandfather Ed.

I moved on to Andrew's file, hoping it would explain more.[15] It turned out that he, too, had once claimed his own homestead in North Dakota, three years before his marriage to Anna. In 1902, Andrew filed a homestead claim near Stanley, North Dakota, a town sixty miles west of Minot. It appeared he had lived in the area for some time, because he filed his naturalization paperwork in the county courthouse there in 1898—he started the process of naturalization in Minnesota in 1893. Instead of waiting out the homestead claim for five years, Andrew paid the government for it a few years early. People could and often did this, in a process known as commutation. It required that they pay $1.25 an acre to obtain title to the land early. Andrew, who paid fourteen dollars to file the claim in 1902, paid another $199.94 in late 1905 so he could claim the title.

The timing surprised me for how close it was to his marriage to Anna. It was perfectly legal to commute a homestead claim. But it was also a practice that lent itself to rampant real estate speculation, the historian Gilbert Fite wrote in *The Farmers' Frontier*.[16] It got so bad in North Dakota that federal inspectors traveled to Minot to investigate. Most homesteads after 1898 were settled by "petty speculators who had no intention of farming," Fite wrote, in a chapter about North Dakota's role at the end of the U.S. farming frontier.

Were Anna and Andrew petty speculators? I created a timeline:

· Andrew applied to buy out his homestead on October 6, 1905.

· Anna got permission to choose another homestead on November 2, 1905. It was one that was in the same county as her brother and one of her sisters, as well as others from her hometown.

· The government approved Andrew's buyout—or commutation—of his homestead on November 25, 1905. He sold the land to another farmer for $3,500, paid to him in $700 installments over five years with 8 percent interest, according to court records I found.

· Anna and Andrew married on December 13, 1905.

· In July 1906, Andrew and Anna officially took possession of her homestead near Larson, the one with the future mineral rights.

There was additional circumstantial evidence in the file. Anna's affidavit requesting a different homestead was on the letterhead of James Johnson, an attorney based in Minot. The same lawyer's name was on the legal notice Andrew was required to place in the newspaper when he commuted his homestead. Johnson took care of the paperwork to make sure Anna could file the claim retroactively as a single woman. Andrew claimed a homestead and then sold it. Then, through his marriage to Anna, he obtained access to another 160-acre parcel of nearly free government land.

It wasn't as though Anna and Andrew got married and then went straight to the county recorder's office. Not exactly. But even if they weren't outright speculators, it appeared there was, at a minimum, some mild collusion with Johnson to ensure Anna snagged a homestead while she still could. And someone, probably Andrew, had to pay the legal bills. But with the money Andrew made selling his original claim, he had plenty of capital to begin married life on a new homestead, the one Anna claimed as a single woman before her marriage. It was savvy real estate speculation, even if it skirted the intent of the Homestead Act, which was to provide land to people who otherwise had no access to it.

The same week Anna and Andrew married, the *Ward County Independent* ran a notice warning off speculators.[17]

> *It has come to the notice of the General Land Office that many homesteaders after proving up their claims sold their land and returned to other states. The government considers that these people have acted in very bad faith, and the special agents have recommended that when such instances arise, the proofs shall be cancelled. Such cases have been known in Ward County. The homesteader who proves up, should by no means place his land on the real estate market.*

The newspapers reported often on General Land Office decisions to cancel people's homestead claims when they failed to comply with the residency requirements of the Homestead Act. A news item shared in a 1904 edition of the *Fargo Forum* told of John Svan of Bowbells, who had no qualms about snitching on neighbors who weren't living on their claims. Such claims were ripe for challenge by people who saw prime land going uncultivated, the paper reported:[18]

> *Svan says that quite a number of young men—and some young women as well—have placed homestead filings on lands over there during the past two years, a large percentage of whom have never made a dollar's worth of improvements on their lands, and several who filed a year or more ago, have never been seen in the vicinity since making their filings. They should not be allowed to cover up these lands and thus shut out honest settlers who would come to aid in settling up the country and improve the rich acres now lying idle and clouded by the fraudulent filing put on by these adventurers.*

I dug up the government report mentioned in Fite's book, the dully named *Report of the Public Lands Commission, 1904.*[19] Its name belied the contents, which were anything but dull. In the report was a rollicking account of the rampant land speculation happening around Minot right at the time Anna and Andrew got married there. It showed that in 1904, people commuted 2,756 homesteads at the land office in Minot. The same office recorded 293 final entries from homesteaders who lived on their land for five years, the path to land ownership anticipated by the law.

The report suggested that 90 percent of the people filing homestead claims were speculators. Some of the commissioners who approved homestead commutations and proofs also sat on the boards of local land companies and banks, the report found. People borrowed money from the bank to commute a claim, then turned around and sold the land for a profit to the same financial institution that gave them a mortgage.

The 1904 report singled out unmarried women as being notorious for speculating on homestead claims. It was especially suspicious of schoolteachers. It seemed downright unfair to single out women, when there was blatant corruption among the men who controlled the capital. But women like Clara Kress must have been all-too-familiar with the contradictions of womanhood of the time: praise for their independence and grit even as the society they lived in tore them apart for daring to reach for an advantage deemed beyond their station.

I don't know why Anna originally filed for a homestead on the other side of the state, nearer to where she was living in Fargo. There's nothing in the file suggesting her motivation was anything other than the land fever sweeping the nation, amplified by enticements in the weekly advertising placed by her lawyer, Henry Krogh and others: "Choice homestead locations—if taken at once." It's also possible that once Anna made her claim, she really did learn the land was unsuitable for cultivation or grazing, as she said in her affidavit. Maybe unscrupulous brokers swindled her, as she claimed.

Or maybe she fell in love with Andrew and decided to file a homestead claim somewhere else with him. It could also be that Andrew was a calculated business decision. Perhaps at twenty-seven, marriage to Andrew was a bargain Anna was willing to make for respectability and stability. Or perhaps it was the pull of familiarity. People she knew from her hometown in Minnesota were making their way to the far northwestern corner of North Dakota.

After reading the *Report of the Public Lands Commission*, though, I developed a new theory. I wondered if the unmarried daughters of some of these Elizabeth settlers weren't the equivalent of a modern-day shell corporation. Their parents settled homesteads in Minnesota. By the early 1900s they owned established farms with value and cash income. Their families might have been able to capitalize the next generation. They might have even sold off their original homesteads in

Minnesota to finance their move to North Dakota. Their adult daughters and sons could apply for homesteads now, too, in North Dakota. Single women like Anna could file for a homestead, and then marry men who had already claimed and sold their own homesteads for profit, like Andrew.

Maybe it was a cynical view. Technically, what Andrew and Anna did was not double dipping. Settlers like my great-grandparents weren't doing anything illegal. They were modest people taking advantage of a liberal interpretation of the law to engage in the time-honored American tradition of real estate speculation.

In recent years, scholars have challenged some of the assumptions in the early homesteading analysis, especially around fraud and the role of single women homesteaders. Recent scholarship on the Homestead Act suggests that fraud was not nearly as rampant as outlined in the sensational reports of the early 1900s. What was considered fraud was more likely an intentional investment strategy used by some poor settlers "to accumulate sufficient capital to establish viable farms," the authors of *Homesteading the Plains* wrote in a 2017 reexamination of the Homestead Act's impact on American life.[20]

What the Homestead Act did represent for many white Americans, though, was generational wealth at its start, funded by nearly free land grants from the federal government. The federal government displaced Indigenous people, often with violence, and then sold their territory at a discount to settlers, including my immigrant great-grandfather and his American wife. Now, more than a century later, there was potential oil wealth on the land Anna claimed. If an oil company drilled, my family might, once again, profit off mineral rights under land obtained in 1905 by skirting the edge of the law.

I went back to what I knew for certain from the paperwork at the Archives: Anna lived with her sister as a seamstress in Fargo until sometime in 1905. She married in Minot in late 1905, shortly after a flurry of court filings and legal fees allowed her to apply for a new claim. Soon after her marriage, Anna got pregnant. She and her husband moved to their homestead in July 1906. They broke seventy-seven acres of land that summer. In late 1906 they had a baby, my grandfather, Ed.

And then just a few months after Ed's birth, something went awry, something not fully explained in the files at the Archives: in early 1907, Anna took sick.

The files showed that in 1912, Andrew applied to the federal government on Anna's behalf for the title to the homestead. Four neighbors filed affidavits in support of his claim. He had to pull together a sheaf of paperwork to claim title to the land homesteaded in Anna's name. First, the government required that he show he was married to Anna, whose name was on the homestead, not his. The file contained a copy of their marriage application and license and the naturalization paperwork proving Andrew was a U.S. citizen. The file also detailed the work Andrew did to prove up a homestead. The application describes two farmhouses on the land, both small, but valued at $300. One was 8 x 10 feet, the other 14 x 16. The barn, which measured 14 x 30, was valued at $80. A granary was valued at $125. A 20-foot deep well was valued at $15. Five acres of fencing was valued at $25, but the greatest value of the farm was in breaking the virgin prairie. That effort was valued at $480. Picking out rocks on all of that acreage added $100 in value. In total, Andrew made $1,125 in improvements, roughly equivalent to $30,000 in modern money. By 1912, Andrew described 120 of the 160 acres as "all good land for farming." He farmed flax and wheat, according to his file.

In his testimony, the government asked Andrew whether anyone had ever been absent from the homestead since establishing residence.

It is here where he and the neighbors described what happened to Anna. "In 1907 she took sick and is now at Jamestown, N. Dak at the hospital," Andrew wrote.

Charles Ely, a thirty-six-year-old neighbor was the most blunt: "Absent since she took sick in 1907. She is insane. The land has been cultivated by her husband since that time."

"She made good residence until taken to asylum," he added.

Then, in the saddest document I encountered in Anna's file, I saw that Andrew had to go to court to claim legal guardianship of his wife so that he could finish proving up the homestead claim made in her name. His petition for guardianship was filed on February 14, 1912—Valentine's Day. There's no supporting documentation for the petition, just the final order from a judge that year declaring Anna "an incompetent person." The file has no evidence of love between husband and wife, just the sad paper trail of a woman sent to Jamestown and that of a man who continued to farm the homestead claim his wife made before their marriage.

As I left the research room that first day, an attendant looked over my copies, made on sky blue paper to distinguish them from original documents. The attendant placed the copies in a green, locked pouch. I retrieved my backpack, and on the way out, a guard unlocked the pouch and handed me the copies.

Now I needed to go to Jamestown.

7

MOTHERLESS

October 2013

North Dakota crude: *$90.16 per barrel*

THE FASTER I UNSPOOLED THE NEWS FROM THE PAST, THE LOUDER THE microfilm machine whirred. It didn't take long to be swallowed up by the stories scrolling by on the screen. I spent the afternoon absorbed in lurid newspaper accounts of children swept away by swollen spring rivers, desperate lovers caught up in murder-suicide pacts, farm hands surviving disfiguring accidents, and other dramas of misfortune, passion, and bad luck that marked life in an agricultural community with an unforgiving climate.

Two months after my first visit to North Dakota, I was back in the Midwest, in search of anything that would tell me about Anna's girlhood. I now knew she ended up at age twenty-nine in the insane asylum in Jamestown, but I wanted to understand why. This time, I started my search at the Otter Tail Historical Society in Fergus Falls, the western Minnesota town near where Anna was born.

Otter Tail County takes its past seriously, maybe because its white settlers, like those in many places in the West during the pioneer period, were so intent on civilizing what they thought of as an uninhabited wilderness without a history. They needed their own history as evidence they were there to stay. Or perhaps to demonstrate that their sacrifices meant something, or to justify their arrival at the expense of those they shoved aside.

Among the holdings at the archive was a two-volume history of Otter Tail County, edited in 1916 by John Mason.[1] Two volumes! Mason, it seemed, was thorough but skeptical of narrative flourish. "No attempt has been made to write a historical romance," he wrote in the preface. "Many things have been presented and urged which would make a readable story, but have, in every instance, been declined where their only basis appeared to rest in the imagination of the informant."

The Otter Tail County Museum and archives opened in 1972 in a low building with a corrugated, cream-colored exterior. Outside, separating the parking lot from the property, is a short fence with metalwork otters, a common decorative motif in the county. A forty-foot otter statue named Otto sits overlooking a lake in a city park on the edge of town. (Of course I took a selfie there, with Otto looking over my shoulder.)

Inside, an archivist asked me to fill out a form detailing the purpose of my research. I checked off "book," "article," and "curious," all in pencil. She scolded me twice for using an indelible pen, not a pencil, to scrawl in my notebook. "Pencils please!" she said. This was a serious archive, her tone said, with valuable historical records. Just like the National Archives.

Reluctantly, I also checked the "genealogy" box. I wanted to distinguish myself from the retirees who, on a vibrant fall day, sat inside a windowless, fluorescent-lit room indexing obituaries and cemetery headstones. Their efforts reminded me of one of my father's many aunts, a woman who, when I was a teenager, shared typeset-and-bound

copies of her family tree research with our extended clan. She claimed
to have traced the Norwegian side of my dad's family back to Erik the
Red. I flipped through the pages of her research in the backseat of my
dad's minivan on the way home from a family gathering in Montana.
I was doubtful of its accuracy. It was clear to me, even as a teenager,
that we came from a bunch of modest farmers, not an explorer who
colonized Greenland.

Genealogy, I thought, was for people who realized late in life that
their mark on the world was slight. Or for members of the Church of
Latter Day Saints who wanted to baptize the souls of their ancestors.
It was, I thought, for those who needed to define themselves with the
stories or the accomplishments of the people who came before them,
instead of living their own lives. No one ever wanted to admit they
were ordinary, or own up to the fact that their ancestors were part of a
government-sponsored real estate grab, one that drove off by force the
rightful inhabitants of the land.

Yet there I was, also cooped up in that windowless room on a beautiful day, delving into a past compiled mostly by volunteers, in order to
make sense of my own life. I had even quit my job to do it.

No matter what we're looking for, the compulsion to understand
where we come from is universal. We live and breathe stories our
whole lives. We need to construct them to make sense of the world we
inhabit, to survive. I wasn't that different than the great-aunt in search
of my family's roots. The only distinction was that I was actively in
search of the dark stories, the ones about the land-grabbing, nuthouse-
residing ancestors no one remembered or wanted to claim.

Anna's family settled on a farm in Elizabeth, a small town twelve
miles north of Fergus Falls, in the rich farming country of the Red River
Valley. Elizabeth was a tiny community, but two years after Anna's birth
in 1878, a railroad spur connected the town with Fergus Falls. Both

the Northern Pacific and Great Northern Railroads stopped in Fergus Falls. As a young girl, Anna could have traveled nearly anywhere in America on a train, beginning with Fergus Falls. It was remote, yes, but very quickly, the scheduled arrivals and departures made it much less of a frontier for those who could afford the train fare.[2]

The historical society had files on many of its original white settlers, but nothing on Anna or her family. The archivist couldn't find anything about Anna's mother in their obituary index. There was also nothing about my ancestors in the township histories the archivist pulled from boxes stored in a back room. Other people with the same last name as Anna had notable roles in the community, but the Sletvolds who merited newspaper clipping files didn't appear to be close relations.

This wasn't unusual. Many Norwegian immigrants to America shared surnames that reflected the name of the farm they came from in their home country, not necessarily any family connection. It could make for frustrating searches in public records and genealogy databases. People also often spelled their names differently in the Norwegian church records than they did in English. Also confusing: Many of the young women from Norwegian families went by their middle names, not their first names. And then when they got married, they went by their husband's names, of course, making it even more difficult to trace them.

Anna's parents were among a wave of Norwegian immigrants who arrived in western Minnesota in the 1860s and 1870s. Crop failures and stratified social conditions in rural Scandinavia in the 1860s spurred the mass immigration from Norway. Federal policy in America also proved welcoming. Thousands of people, not just Norwegian immigrants, arrived in Minnesota after President Lincoln signed into law the Homestead Act in 1862. Anna's father arrived in the United

States in 1866. By 1870, there were about 50,000 Norwegians living in Minnesota; by 1880, there were more than 120,000.[3]

Many of those Norwegians made their way to the farmland around Fergus Falls after the U.S.-Dakota War of 1862 resulted in the mass hanging of thirty-eight Dakota in Mankato, Minnesota.[4] Subsequently, the federal government forced thousands of Dakota to internment camps. Military leaders had no mercy for the Dakota, who warred with civilian settlers after government officials refused to distribute food to those who were starving. "It is my purpose utterly to exterminate the Sioux if I have the power to do so and even if it requires a campaign lasting the whole of next year," U.S. General John Pope wrote to Henry Sibley, a colonel in the Minnesota state military forces. "Destroy everything belonging to them and force them out to the plains, unless, as I suggest, you can capture them. They are to be treated as maniacs or wild beasts, and by no means as people with whom treaties or compromises can be made."[5] Whether from forced removal or death by starvation, by 1867, historians estimate that only fifty Dakota remained in Minnesota.

The war made land available in Minnesota for white settlers like Anna's parents, free from the perceived threat of the Dakota people. The new Norwegian arrivals worked hard to build familiar communities not unlike the ones they left behind in Norway. Church services were in Norwegian, and Anna's mother and father could read a weekly newspaper, the *Ugeblad*, in their native language, beginning in 1882. Even the landscape and the climate were familiar. In the time before lumberjacks felled the trees, there were extensive pine and hardwood forests. It was a green, verdant place in summer, and very cold in the long winters.

Yet it wasn't the easy life many of the immigrants were promised back in Norway. I found in that two-volume county history book an account of one of the grasshopper plagues that descended on wide swaths of Minnesota in the 1870s. A man named Ole Jorgens, who was the county auditor, remembered the pests raining down on his horse and buggy in 1873 as he drove through Elizabeth.

> I soon noticed that they had begun their deadly work—the grain in the fields, the vegetation in the gardens, all things green along the roadside showed the effects of their rapacious appetite. Hardly a head of wheat was left on the stalk and the garden vegetables were not only gnawed off to the ground, but in many cases even the roots had disappeared. In the case of even onions the hoppers had not only eaten the stalks, but even the very root of the onion, so that nothing was left of it except a hole in the ground where it grew.[6]

Anna's parents, Edward and Martha, arrived in Otter Tail County in the midst of this biblical plague. Anna wasn't yet born, but her older sister, Christine Marie, was an infant. Christine's arrival coincided with another scourge: the Financial Panic of 1873. The Northern Pacific Railroad, based in Minnesota, declared bankruptcy, triggering a widespread economic crisis that led to a national, multiyear depression.[7]

All of Edward and Martha's children arrived during that depression. Anna was the second oldest of at least six, their names stairstepping predictably through twenty years of state and federal census forms in the late 1890s and early 1900s.

So far in my research, though, their mother was a mystery. Martha shows up in an 1885 state census of Minnesota. She is thirty-two; her husband Edward is thirty-nine. Anna's older sister, Christine, is twelve. Anna is seven. She's followed by Ole, five, Edward, three, and baby Olava, who was one.

But then Martha's name stops appearing in the census records with the rest of her family. In the 1895 Minnesota census, it's just her husband, followed by Anna, who is seventeen. Next is Ole, fifteen, Edward, twelve, Olava, eleven, and a new entry: Etta, nine, who was born just after the 1885 census. (Anna's oldest sister, Christine, was long gone, married, and living in Fargo, according to my search.)

Martha seems to just...disappear. I couldn't reconcile her absence with something as bureaucratic and linear as the once-a-decade accounting of people's whereabouts. Where did the mother of five young children go between 1885 and 1895? The 1890 federal census was no help, not to me or to anyone else looking up their ancestors. Most of the forms were destroyed in a 1921 fire at the Commerce Department building in Washington, DC.

I had a hunch that Martha died when Anna was young, but I couldn't find a record of her death, not online and not at the historical society. I drove over to the Otter Tail County courthouse to look through the books that once recorded every birth and death in the county. These records were also at the historical society, but as a printed index of names. Since I was there in Otter Tail County, I wanted to see the original documents for myself. I knew it was possible to make unexpected discoveries simply by flipping to the next page of an old record book.

At the courthouse, I paged through the ledgers from 1885 to 1895 in search of evidence of Martha's death. I didn't find it. Instead, while looking for her name, I unearthed something much more unsettling.

There, in the birth-and-death ledger, I discovered that Anna had two even younger siblings: twins Minnie and Theodore. They were born to Martha Sletvold on January 11, 1887, when Anna was eight.

Excitedly, I texted my sister, the mother of twin girls herself: "Twins do run in the family!"

Minnie and Theodore were cute names. I entered them in my phone, in a secret file where I kept a list of the potential names of my future children. The list got its start one day when Chris and I walked together through Rock Creek Cemetery, a graveyard near our home in Washington, DC. We were captivated by old-fashioned names on the graves, like Cassia and Calvin. Minnie and Theodore were equally charming additions to our list.

But a few moments after that text, I was crushed: Minnie and Theodore showed up in the death index just months after their birth. Minnie was three months and sixteen days old. Theodore lived a little longer; he was four months and six days old when he died.

I can only speculate what caused their deaths; the index didn't have that information, and there were no death certificates for them. It could have been diphtheria, a scourge of childhood at the time, but it seemed unlikely. The Sletvolds had six children, one of whom, Etta, was barely a year old. The other children would have been afflicted, too. It could have been malnutrition, although that also seemed unlikely. They were not rich, but I had no evidence the Sletvolds lived in abject poverty; life may have been difficult, but they owned a farm and paid a mortgage and went to church. They all knew how to read and write and speak English, according to the census forms.

What seemed most likely is that the twins were born too early, with small lungs that challenged the likelihood of their survival. Maybe Minnie's name was due to her diminutive size.

I left the courthouse, a low, modern brick building adjacent to the grounds of the Fergus Falls State Hospital. The former asylum was abandoned despite the city's best efforts to redevelop the Victorian structure. I wandered the grounds for a few minutes, reluctant to go back inside the archives on such a beautiful day. It was a lovely place in fall. Leafy hardwoods blazed with color. Maybe, I thought, old newspapers would have more information about Minnie and Theodore.

I returned to the archives, whirring my way through the old newspapers. They had a low threshold for what they deemed newsworthy.

Everyone's business appeared to be on display, no matter how small the matter. The newspapers reported on the comings and goings of people embarking on or returning from long journeys—or even journeys a few towns away. If someone was sick, you read about it in the paper. Just like the North Dakota papers twenty years later, they also made note of who was sent to the state insane asylums. I was certain that the death of young twins in that spring of 1887 would have been noted in the *Fergus Falls Weekly Journal,* which included short dispatches from nearby townships, including Elizabeth. Sent in by correspondents, the items varied widely in quality and depth of coverage. Sometimes, weeks went by in the summer without updates, one of the archivists in Fergus Falls told me. People who reported the news were too busy on their farms to write up and send in what they'd heard.

As hungry as I was for a mention of Theodore and Minnie, I couldn't stop reading the nosy news reports. There was a clear line of progression from the pioneer newspapers to the stories modern journalists report. Just like the past, modern news thrives on weather reports, political squabbles, and violent death.

Using a pencil, of course, I scribbled loose, poorly annotated notes of the cacophony of life in and around Fergus Falls in the winter of 1887, during the span of Theodore and Minnie's short life.

In February 1887, I read, trains in the county ground to a halt because of severe weather. Elsie Josephson, described in the paper as insane, "kills baby and self." The power company installed light poles; one hundred merchants in Fergus Falls signed up for one-month trials. Levi Bros. was the first to install electric lights. Farmers killed twenty-three hundred gophers with poison. Dr. Charest of Perham was described as the township's first owner of a bicycle. All Otter Tail County Indians were "ordered onto reservations," the paper reported dispassionately. "Mounted police to corral and move them." David McCollum, thirteen, lost his index finger at the Red Eye sawmill. Another grim story capti-vated me: the murder of fifteen-year-old Lillie Field. A man named Nels Olson Holong was accused of mutilating her with a knife "because she refused his attention." The paper reported that he tossed Lillie's body into a pig pen, "where it was partially eaten."[8]

Nobody wanted to claim that guy on their family tree.

The biggest news in the *Fergus Falls Weekly Journal*, though, was about the state hospital for the insane, under construction on a hill on the north side of town. The state budgeted $240,000 for the facility, which would have room for six hundred patients when it opened. It would also relieve overcrowding at the asylum in St. Peter, the other closest mental hospital.

Then, I found it. One short mention of the twins, but not by name. It was a tiny little item in the January 20, 1887, edition of the *Fergus Falls Weekly Journal*, a newspaper not available in any online index at the time: "Ed Sletvold is twice glad. A little boy came into his house last week and a little girl too."[9]

"Twice glad." It was a beautiful phrase that almost made up for mentioning a little girl as an afterthought. At least, I thought, the family had some measure of joy, however brief.

I kept searching for any mention of their death, but I turned up nothing. It wasn't until later, when I plugged their last name into a dig-itized newspaper archive run by the Library of Congress that I found another clue. It was in the May 11, 1887 edition of the *Ugeblad*, the weekly Norwegian-language newspaper.[10]

I copied the text, and then pasted it into an online translator:

It wasn't quite right when we announced last week that there was a child dead for Ole Sletvold. It was Even Olsen Sletvold. Now, however, another of his children is dead. They were buried Saturday. The two children were twins some months old.

My stomach dropped. It was a correction. Every journalist makes mistakes. They are usually small, unintentional errors or typos, but when you make one you feel as though it is the end of the world. And you dread making them. I couldn't count the times I awoke at 3 a.m., convinced I spelled someone's name wrong. Or that I wrote millions when I meant billions. Or that I accidentally identified someone as a Democrat who was a Republican.

From my research, I understood how the mix-up could happen. Anna's father went by several names, depending on the circumstances. It was "Even" in many of the church records and "Edward" or "Ed" or "E.O." in English. Plus, there was another Sletvold family in Otter Tail County with a similar passel of children, some who even had the same names as Anna's siblings. If it was a tangle in the historical record of the future, it must have been confusing in 1887, too.

But this error of identity was the worst sort of error, the kind where a journalist gets something wrong that has the potential to cause great emotional harm to a family. This was a bad mistake. I could imagine how awful the editor felt when people alerted him to the error. I knew the burn of shame in my gut over getting something wrong.

As terrible as the correction was, the news for the family got even worse. One column over, the newspaper had a small item that explained Martha's absence: *"Martha C. Sletvold, sent last Monday to St. Peter as insane. She has also been there before."*[11]

The newspaper editor failed to understand that these were interconnected stories. The woman who gave birth to the dead babies one column over was also the Martha Sletvold who was sent to St. Peter, then the nearest insane asylum to Fergus Falls.

One particularly sad line stood out to me: "She has also been there before." It meant that Anna and her siblings were, essentially, motherless. It meant that Martha's children already lived with the uncertainty that their mother might go away at any time. And yet Anna was only eight years old when all of this happened. Surely, she had some time to be a child before she and her older sister had to mother their four younger siblings?

I thought of my life at eight. Eight was when my friends and I learned to hitch a leg over the horizontal bars on the playground, one femur in front of the other. We wrapped our hands around the metal, but we barely needed to hold on. We twirled, anchored by the centrifuge of our spinning. We spun and spun, dizzy princesses, until it was time to let go. Then, we landed, wobbly and woozy in the barkdust. Eight was a birthday picnic with friends at the beach, with a bonfire and hotdogs and s'mores and kites and sandy feet. Eight was ice cream sandwiches in sticky paper wrappers, bought from the vending machine after swim lessons in the overly chlorinated municipal pool that turned my sister's blond hair green. It was pigtails and a fascination with unicorns and rainbows. It was at eight that I began reading on my own the very books that gave me any knowledge of a Midwest girlhood in the 1880s.

I tried to imagine Anna at eight, the summer before it all went wrong for her family, the summer before the birth and death of her baby brother and sister. No matter what, to be an eight-year-old girl on a summer day on a farm in Minnesota in 1886 would have been just as magical as it was to be eight in rural Oregon in 1981. I imagined Anna sneaking away to wade in the Pelican River. Maybe she and her brothers caught fish or chased butterflies. There were four ice cream parlors in Fergus Falls at the time, according to my research. Surely Anna got to go to town for ice cream at least once? I could see Anna and her young brothers on their backs in the grass, watching as the clouds built into afternoon thunderstorms. Their fingers were stained with berry juice, their mouths smeared with the wildness of an unsupervised summer.

Yet again, I was romanticizing Anna's life. She had brothers ages six and four, a sister who was barely two, and another who was an infant. It's unlikely Anna had any attention at all from her mother, occupied as Martha was with a baby girl on her hip and pregnant with the next two little ones. As one of the oldest children, Anna would have been required to mind the younger ones. Her sister Christine would have had even more responsibility.

The entire pioneer period "was kind of a children's disaster," wrote David Laskin in *The Children's Blizzard*, an account of the January 1888 storm that killed several hundred people, many of them children who froze to death on their way home from school in the Dakotas, Nebraska, Iowa, and Minnesota. There was collective, national mourning over their deaths, Laskin wrote, because people knew how hard children had it on the Great Plains. Mourning was all they could do for these children without childhoods. Children were "the unpaid workforce of the prairie," Laskin wrote. "A safe and carefree childhood was a luxury the pioneer prairie could not afford."[12]

Martha was "sent to St. Peter," just days before Anna's ninth birthday. Anna at nine would have been a different child than Anna at eight, marked as she was by the loss of her twin siblings and her mother.

Yet even such losses may not have prepared Anna for what happened next. Five years after Martha went to St. Peter, the *Ugeblad* once again reported on a Sletvold family tragedy. The article ran on the back page of the September 7, 1892 issue of the weekly newspaper.[13] Again, I ran the Norwegian through an online translation program:

> *It's a shame and for the family a very sad story from St. Peter's Psychiatric Hospital. Mrs. Sletvold of the town of Elizabeth in this county is one of the patients there and has been there twice. Last May she gave birth to a child, but the matter has been kept so secret that newspapers have not heard of it until now.*

The case has been investigated in the hospital, but it has not been possible to know who is the father of the child. Mrs. Sletvold refuses to provide any information about it. Mr. Sletvold and his wife's brother were informed of the affair, but neither one got anything out of her, it is said.

Mrs. Sletvold's sense must be completely gone, but since she has been very calm and has done nothing wrong, she has enjoyed some more freedom than some others, and it seems someone took advantage of her. There are about 500 female patients at the hospital. They are in a separate part of the building with female guardians and a woman doctor.

The account in the *Ugeblad* stunned me with its callousness. It was a gossipy, carelessly wrought story, which was a different, more deliberate sort of sin in journalism than identifying the wrong man as the father of dead twins. The article took pleasure in unveiling a private family matter, when its author should have been concerned with righting the injustice contained within the secret.

The article was written in 1892, at a time when American women had growing but still limited legal, social, or political power. But even for the era, it was terrible framing. Instead of casting the blame on the unknown man who impregnated Martha, the newspaper recounted the events as a titillating scandal. Instead of describing what happened as a crime or as a lapse in institutional oversight, Martha and her family bore the blame and the shame of the crime.

What happened to Martha wasn't just idle gossip. No woman in an institution such as St. Peter had full agency over her body, and because Martha was there against her will, no sexual relationship she entered into at the asylum could have been truly consensual. The pregnancy was the result of a sexual assault on a thirty-eight-year-old woman made vulnerable by her involuntary commitment.

Every single Norwegian-speaking member of their community

must have known of the scandal. Anna was fourteen when it happened. The next year, she was confirmed in the Lutheran church in her community. She had to pledge to live a life of faith in front of an entire congregation of people who knew what happened to her mother.

I found only one mention of the baby in the historical record: a child born in May 1892 to Martha Sletvold was baptized at a Lutheran church in St. Peter the next spring. Her name when she was baptized was Ida Christine, although I suspect the baby was adopted and renamed soon after, because it was what was done at the time. I could only hope the baby was welcomed into a loving home with parents who wanted, but who could not have, a child of their own.

Martha Sletvold died of heart failure in 1918 in Anoka, another asylum in Minnesota. Her exact birthday is not listed on the death certificate, and the document said she was sixty-two; previous records I found suggested she was sixty-five. Martha's death certificate had few details other than this handwritten note: *"She died suddenly."*

Unlike the thrill of discovery I experienced when I unearthed Anna's story from the files at the National Archives, the revelations about Martha left me with a sense of deep sadness and resignation. Anna and her brothers and sisters lost their mother to the asylums long before Martha died in her early sixties of heart failure. *She has also been there before.*

As I drove away from Fergus Falls, I considered not only the end of Anna's innocence but the end of my own. At nine, a classmate of mine playing unsupervised in the Yamhill River drowned after school in the fast-moving, spring waters. Jason was the class troublemaker with few friends, the only son of a single mom. But he was goofy and lovable with his eye-grazing dirty blond bangs and filthy jeans. His teachers always looked out for him because they could see he needed so much. Mrs. Johnson, our teacher, cried at Jason's funeral. It was the first time

I noticed an adult in authority weeping from sorrow in a public place. After the funeral, my mother took me and some classmates to a café across the street from the church for sodas, so we could talk if we needed to. We didn't have grief counselors then, not in our little town. But I learned something from my mother that day: sadness needs space, too.

Until Fergus Falls, I hadn't thought about Jason in many years. I couldn't remember his last name, which felt cruel. I wish I had been kinder to Jason. Nine was different from eight.

This was difficult and lonely research. There was no immediate gratification, like with a news article that went online as soon as you had enough information to publish a complete report, or even one that printed the next morning in the newspaper. Instead, when I unearthed something from the past, I followed it as far as it went, never quite sure how it fit into the overall narrative of Anna's story. It was intriguing, and the reporter in me enjoyed the search. But it also meant spending all of my time in the land of the dead, in a place of sorrow and regret and loss. It meant long days in the car, with nothing but my thoughts for company. Sadness had almost too much space.

I missed my mother, who would have been good at the genealogy search, had she been alive. She would have liked traveling beside me in the car from Fergus Falls to Jamestown, recalling the snippets and whispers of stories she heard as a child. My mother would have been delighted with the intrigue. She would have seen the drama in the stories I unearthed, and even though she wasn't a professional writer, she would have understood how to amp it up for narrative panache. She would have helped me understand what I had uncovered.

Through the generations, I could feel the reverberations of the Sletvold family's shame and loss. It was a vast, echoey absence that seemed embedded in my very genetic code.

I drove west to Jamestown. Motherless, like Anna.

8

ANNA

October 2013

North Dakota crude: *$90.16 per barrel*

ANNA IS BURIED IN A SMALL CEMETERY ATOP A BLUFF ABOUT A MILE FROM the North Dakota State Hospital in Jamestown. Her grave marker sits flush with the earth, a rose quartz rectangle bordered in gray marble. There's a quiet hum from a nearby electrical substation and the cars on the interstate and, in late October before the first hard frost, the buzz of the last of the cicadas. Someone, most likely inmates, planted trees so long ago that they're now tall enough to provide shade and some shelter from the winds that blow unimpeded from Canada. Western meadowlarks sit atop the fence posts, singing cheerfully in an otherwise somber, rarely visited place.

All graves are marked, including Anna's. The state gave the hospital a grant to replace older, oval-shaped headstones with marble grave markers. But the hospital ran out of money to complete the work, and the cemetery remains a mixture of the older and newer styles. Some of the soft gray stones on the older graves are so worn from the North Dakota winters that names are barely legible. The rest sit flush with the earth, like Anna's.

Anna's grave in Jamestown, North Dakota.

I wasn't expecting to find or visit Anna's grave when I reached out to the hospital in search of her medical records; I was just looking for a file. Anyone who enters a state institution has some sort of file, and even though I knew it was a long shot after ninety years, I figured it was worth asking.

My reporter's hunch was correct: Anna, it turned out, did have a file. I couldn't access it as a journalist, not without a fight, but the hospital gave it to me as one of Anna's descendants. I asked at just the right time. The North Dakota Legislature had recently passed a law allowing hospitals to destroy medical records after seven years. Anna's, somehow, still existed.[1]

I picked up copies of Anna's records in the office of Karla Bachmeier, a state hospital employee. Bachmeier handed me a fifteen-page sheaf of photocopies and told me she would show me where my great-grandmother was buried.

She told me she arrived at work early that morning to find and clean Anna's grave for me. I was grateful but not sure how to accept this unexpected kindness toward a distant relative of a former inmate. Anna was my great-grandmother, yes, but no one I knew had ever known her, her own son included.

I followed Bachmeier's minivan to the gates of the cemetery. Once she drove away, I sat on the dry grass near Anna's grave to read her paperwork. The day was bright with golden autumn light, clear blue skies, and the sort of mid-70s temperatures the weather reports kept calling "mild" for October in North Dakota.

———

A judge declared Anna insane on April 5, 1907, in the same courthouse in Minot where she and Andrew applied for a marriage license just sixteen months before.[2] All of this was detailed in a written report by Dr. S. G. White, who oversaw Anna's care and supplied the supporting evidence for her insanity declaration.

Anna spent March of 1907 confined to the local hospital in Minot. Anna wasn't suicidal, the file noted, and she wasn't addicted to alcohol, opium, tobacco, or "intemperance in any form." She also wasn't disposed to "filthy habits, destruction of clothing, breaking glass, etc."

But beginning in January 1907, shortly after her son's birth on December 29, Anna was "melancholic" and had "no regard for her child," Dr. White reported. She was sleepless and had a poor appetite. She had delusions and thought that people were trying to harm her. The symptoms were only getting worse, the paperwork noted. Anna "made threats to injure several people. Has even struck them."

Has the patient been subject to any bodily disease—epilepsy, suppressed eruption, discharge of sores or ever had any injury of the head? *Following confinement.*

Has restraint or confinement been employed? *Yes. Hospital for 1 month in Minot.*

What is supposed to be the cause of the disease? *Confinement.*

What treatment has been pursued for the relief of the patient— mention particulars and the effects? *Moral restraint. Sedative + tonics. Curettage—no effect.*

I took note of Dr. White's use of the word "confinement," in all its meanings. One use described Anna's literal confinement in the hospital in Minot for a month—her restraint there. Its other use was an old-fashioned euphemism for pregnancy.

More puzzling, though, was the mention of "curettage," a word used most often to describe a procedure that scrapes the uterus of excess tissue, especially after a miscarriage. It suggested that Anna suffered from severe postpartum bleeding from intrauterine tissue she didn't deliver after she gave birth.

Dr. White's terse answers on the form, a bureaucratic formality required by the court, were the only explanation of the months leading up to the judge's order to commit Anna to the asylum. Anna's son Ed, my grandfather, was just three months old at the time. She was having delusions and had "no regard" for her baby. She also likely had uncontrollable post-birth bleeding.

Across the generations, my heart ached. A doctor sent Anna to the mental hospital because she had severe postpartum depression, possibly even psychosis. Modern women have a difficult enough time with this diagnosis. I could only imagine Anna's terror in 1907 as she lost autonomy over her body, with its continual bleeding, and the control of her mind.

She understood what it meant, because she saw it happen to her own mother.

———

Anna arrived at the North Dakota Hospital for the Insane at 10:30 a.m. on April 6, 1907. To get there, she and her husband took the Soo Line train from Minot, and then switched to the Northern Pacific line at Carrington, a farm junction, to travel south to Jamestown. They were accompanied by a relation listed in Anna's hospital admissions paperwork as Mrs. Dayton. Anna's infant did not come with her. Unless Anna was completely out of her mind—she was almost certainly sedated for the journey—she knew exactly where she was going. To Jamestown.

The train journey took them along present-day U.S. 52, through towns named Velva, Voltaire, Harvey, and Fessenden. They passed through the extraordinary terrain of the Missouri Coteau, a diagonal swath of grassy, rolling hills and rocky soils that runs from the north-west corner of the state down to Jamestown. In April in this landscape, snow melt ponds on the thawing earth, creating a habitat for migratory ducks, shorebirds, and songbirds. The seasonal pothole lakes shimmer in the spring light. Large flocks of returning geese fly overhead, so numerous they cast shadows on the earth and water below.

As Anna and Andrew and Mrs. Dayton exited the train in Jamestown, the state hospital loomed above them in the distance, atop a bluff overlooking the James River Valley. From the depot, the unlikely trio crossed over the river, then rode up to the asylum.

The main building, a solid, four-story reddish brick structure with a gabled roofline and a tall central tower, was built in 1885. As the state grew, so did the hospital's population, and so did its campus. During Anna's time there, the population nearly tripled from 515 people to 1,288. The hospital detailed the growth in its biennial reports to the state legislature. Each patient cost nine dollars per year in butter alone, the 1906 report said, in an entreaty to lawmakers to budget for a cold storage unit on the site, and to install electricity to run the laundry.[3] By 1920, it cost $11,000 per year for fire and tornado insurance.[4]

———

On her arrival, Anna was 5 foot 6 inches and weighed 119 pounds, down from her regular weight of 134 pounds, according to the admission notes. Her clothes were of good quality and in good condition.

She had with her two gold rings and a purse containing 79 cents. The diagnosis on the front of her file: *Manic Depressive Insanity*.

Her first twelve days were a new hell:

April 6: *Admitted to Ward 9. She is quite depressed and inclined to resent the fact that she cannot go about where she likes.*

April 7: *Slept well last night but does not "care to stay here."*

April 9: *The patient is anemic.*

April 11: *In talking with this patient she has a grievance about staying here and does not care to take "medicine" or do anything. She has been out of bed for the past two days.*

April 18: *See gynecological examination. It is almost impossible to persuade this patient to go to bed at night and generally the attendants are obliged to undress her. She occasionally refuses to eat for one meal but on the whole has a good appetite.*

There are no notes on the gynecological examination. Her file has no entries for another three months, not until Anna tried to escape.

July 25: *Continues about the same as heretofore. Never wants to undress. On going to bed at night it requires constant urging and coaxing for almost an hour to get her to retire. She is depressed and always wears a sad expression. Is dissatisfied with everything— her food, her bed, her clothes, and everything that is attempted for her comfort she interprets as abuse and ill-treatment. Was transferred to Ward 1 about two weeks ago, in the hopes of making her more cheerful in more cheerful surroundings, but she attempted to escape one evening, and was sent to Ward 2 where she is at present.*

1908: *Scolds for hours at a time; considers that she is very much abused.*

Anna tried to escape several times the first few years. She must have known that if she did not leave the asylum, her fate would be to die there. She was nearly successful just once, on January 30, 1908, after she had been an inmate at the North Dakota State Hospital for ten months.

This morning, while going to the general dining room, she escaped. She was found an hour later at the station at Jamestown. She was returned to Ward 4. The weather was mild so she did not suffer from exposure.

I looked up the weather for that day: the high was twelve degrees. It would have been a severe walk of several miles, presumably without a coat or hat.

Anna would have hurried, anxious to get on a train before she was found out. Her odds of success were low. She had no means to buy a ticket home. She would have had to persuade someone at the station that she would pay when she arrived at her destination. Perhaps she told them: *I have not seen my child in a year.* Perhaps she begged: *Please let me on the train. I will pay you when I arrive home.*

And who would believe her, this wild-eyed woman without a coat or hat in January, her cheeks flushed with cold? Everyone at the train station would have known Anna was an escapee from the asylum. None of the men in charge would have allowed her on the train.

The ride back to the asylum from the station must have felt like defeat. I imagined Anna slumped in the back seat of a buggy or an early automobile. She would have watched as the last of her freedom slipped away, her cheek pressed up against the window as an orderly drove her back up the bluff to the state hospital.

Anna's file offers no answers, but January 30 must have been the day she stopped trying to escape, the day that the indefinite state of detention began to define her life. The organizing structures of life, of adulthood unfolding into parenthood into middle and then old age, would have ceased to define her. Her days would instead be defined by the rhythms of an institution. Communal breakfasts, lunches, and dinners. The smell of disinfectant when custodians swabbed the floors clean. The Victrola in the afternoons. Weekend dances in the winter. Staff baseball games in the summer. The edges of her existence would have been dulled by the tonics they forced on her. Drugged into submission, Anna might not have noticed as days and weeks and then years stood between her and freedom.

In the moments before she woke, in that liminal state before the reality of the day, I hoped she clung to the belief she was free. I imagined her in that first year there as she opened her eyes and said out loud, "Today is the day I leave this hell."

And yet there must have been a day where she stopped saying it, a day where she awoke and understood she would never leave the asylum. If she had access to a calendar, maybe she noted it: "This was the day they broke my spirit."

———

Days turned into weeks, which turned into months and then years. The rest of Anna's file spans fourteen years, with a gap in the file only in 1916, during World War I. The file in its year-by-year summary of Anna's health and behavior shows an unrelenting unraveling, even as the events of the outside world continued without her. World War I began and ended. Women got the vote. Mary Phelps Jacob patented the brassiere. And the population of North Dakota boomed from 190,983 to 646,872 people.

1909: *Always disagreeable and cross. Physical condition good. Always idle.*

1910: *Walks up and down the short hall with her dress draped over her arms, all the while talking to herself. Has delusions of persecution.*

1911: *Careless of her personal appearance; often destroying her clothing and neglecting to comb her hair. Taking tonic three times a day.*

1912: *Mental condition remains about the same. Patient's speech is abusive and profane.*

1913: *Talks to herself, becoming profane and abusive when spoken to. Helps with the ward work and the remainder of the day paces up and down the short hall.*

1914: *No change physically since last note.*

I saw no mention of it in her file, but it's possible that doctors sterilized Anna at the hospital. In 1914, the North Dakota legislature authorized involuntary sterilization of inmates. There was a widespread belief that insanity was heredity. The first year, eleven men underwent vasectomies in North Dakota. By 1918, doctors were using x-ray equipment to sterilize women at the asylum. It remained "a routine procedure at the hospital for decades," wrote James Smorada and Sally Van Beek, in a 1983 history of Jamestown that takes a surprisingly candid look at the failures of the institution looming over the city.[5] Other histories dispute the frequency, saying that only a few people were sterilized in North Dakota during Anna's time.

Nonetheless, the law reflected the belief at the time in the United States and beyond that mental illness was hereditary. This understanding influenced how doctors diagnosed women such as Anna. If the doctors believed Anna's mother was insane, the medical thinking of the time suggested Anna was, too. And because Anna's mother had a secret, scandalous child at an asylum, it is possible Anna was seen as a natural candidate for sterilization.

Anna was confined at the peak of her fertility so she couldn't give birth to children who, in turn, might also be insane. This thinking kept Anna—and women just like her—locked up for the rest of her life.

1915: *Quiet and orderly. Works on the ward. In fair physical condition.*

I found a description of the asylum from 1915, tucked away in the Jamestown file at the North Dakota State Historical Society. It was a relic written by a woman alive at the same time as Anna, a journalist named Mrs. W. F. Cushing, who later ran a newspaper in Fargo with her husband. Mrs. Cushing wrote the pamphlet to extol Jamestown's business and civic virtues: "On the brow of the bluffs overlooking the valley of the James River and to the south of the city, stands the North Dakota Hospital for the Insane. The buildings are ornate and can be seen for many miles in all directions."[6]

What Mrs. Cushing didn't know or didn't write, is that when you get up close to such a place, the view changes. The ornate becomes utilitarian. There's a reason all the insane asylums once sat atop hills, looming over the towns they served. It is not to haunt us, although they do, these brick Victorian structures built to hide people whose minds and habits society deemed too chaotic to live down below in the flats. These places sit up high for a far more practical reason: human waste. Any institution housing hundreds of people needed vast septic fields for all that shit flowing down-hill. The year Anna arrived, the superintendent asked for $5,000 to build a septic system. Before that, all waste emptied directly into the James River. "There has been considerable complaint about this nuisance," the superintendent conceded in his report to the state legislature.[7]

> **1917:** *This patient never speaks to anyone. She ignores the presence of other people. She keeps her person neat and orderly. She walks up and down the short hall in the ward almost all the time during her waking hours. It irritates her to speak to her. She sometimes strikes attendants and other patients.*
>
> **1918:** *Usually conducts herself well, she does not take much interest in her surroundings, her mental condition is below par.*
>
> **1919:** *Assists with the ward work. There is no change in her mental or physical condition.*
>
> **1920:** *Patient is usually quiet and orderly, but at times is rather irritable and stubborn. She is rather stupid and her mental condition does not improve.*

The superintendent's report submitted to the state the year before Anna died suggests there was a small measure of kindness in the hospital's operation. Mental illness, wrote Dr. William Hotchkiss,

> Does not hesitate to strike the proud and affluent, the brilliant and powerful as well as the lonely and poor.
>
> Insanity is one of the unfortunate heritages of mankind and steals upon its unsuspecting victim like a thief in the night. It is very frequently a hereditary misfortune, but often attacks men and women in whom no predisposition has ever been demonstrated and no one can feel perfectly safe and secure, as above and beyond its reach.

By 1920, no restraints were used "except in extreme cases," the report noted, "when a patient is suicidal."[8]

There were 1,200 inmates living at the asylum the year Anna died. It was a small town. The superintendent wrote in his 1922 report to the legislature that music was "a valuable means of diversion and entertainment." An orchestra made up of hospital attendants played at church services and at weekly dances in the winter. The asylum owned a dozen Victrola record players and a library of two thousand records. There were pianos in many of the wards.

An employee baseball team played on the grounds during the summer season. Patients got to attend the circus and the county fair, the report notes, and there was an annual Fourth of July picnic. An extensive library stocked current magazines and books. "Mental cases stay in a hospital for weeks, months, and years; their minds are often alert, their physical condition good, and time hangs heavily on their hands," the administrator wrote. "Many of our mental patients are as much interested in the outside world as anyone. These we supply with the daily papers, the year's calendar, and maps."[9]

The last entry in Anna's medical file is from a few days after her death at age forty-three, from a gallstone infection:

She died April 25, 1921 at 10 p.m. of Biliary Calculi. Her husband, Andrew Haraseth, viewed the remains on the 27th and ordered that she be buried in the Hospital Cemetery which was done on the 29th.

I felt a brief flash of empathy for Andrew, whose return to the hospital to view his wife's remains was yet another sad train journey through remarkable country, once again in April. I envisioned him looking out the window for a moment, lost in regret. Perhaps he took comfort in lakes and skies teeming with birds.

Yet Andrew was unwilling to bring Anna home to be buried near anyone she loved or who might have loved her. Her sister, her brother, her father, and her son still lived in Burke County. Andrew never even told their son, Ed, what really happened. I know this not only from my mother, but because it is also in Anna's file, toward the end.

On June 15, 1947, my grandfather, Ed, wrote to the superintendent of the state hospital. It was after his father's death, and he was asking for information about his mother. My grandfather needed a statement for the probate judge to prove he was entitled to the land Anna homesteaded before his birth. His mother, he wrote, "passed away at your institution about 27 years ago."

"Her name was Anna Josephine Haraseth, wife of Andrew Haraseth," he wrote, in a firm yet loping cursive that crowded the page. "She was committed early in the year 1907 I believe, and passed away in April, in just what year I never was told but believe around 1920."

The superintendent wrote back just a few days later, confirming Anna's date of admission and death. The copy Bachmeier gave me was a poor quality photocopy of the original carbon copy. I squinted in the sun to read the superintendent's typewritten response: "Your mother was buried in our hospital cemetery which is nicely kept and all graves are marked. I trust this is the information you desire."

It was an awkward word, desire. The secretary whose initials were an illegible smudge on the bottom of the page was the one who chose the word. Desire suggests excitement, a natural longing and the pleasure of its fulfillment. For a moment, my inner editor thought "seek" might be the more appropriate verb. Certainly it was more business-like than "desire," especially for a letter that might be entered in a probate file in the Burke County courthouse.

The secretary, working in a place of many unfulfilled longings, might have been deliberate in her word choice. Ed never knew his birth mother. He knew only the legacy of her absence. He was forty years old and halfway through his life when he wrote to the state hospital seeking answers. It took forty years and his father's death to learn what happened to his mother. As a child, Ed lived with a foster mother, Lena, who was still alive when he wrote that letter. But Anna's absence was a presence in my grandfather's entire life, an unfulfilled longing. I reconsidered "desire." Perhaps the secretary chose the right word.

All those Victrola records in the hospital's collection must have included a recording of "All By Myself," a popular Irving Berlin song published the year Anna died. Or maybe someone played it on one of the pianos on the wards. It was a hit. Even in the asylum, Anna must have heard the song.[10]

All by myself I get lonely
Watchin' the clock on the shelf
I'd love to rest my weary head
On somebody's shoulder
I hate to grow older
All by myself.

At age twenty-nine, Anna lost agency of her own body. She, and likely her mother before her, had postpartum depression. Subject to the decisions and whims of men, Anna and her mother had no control over the conditions of their captivity. There was no effective treatment for their severe melancholy, either. Anna's husband committed her to the hospital because he saw in her what he had seen in her mother, Martha, back in Fergus Falls: a woman driven mad by childbirth. A doctor and a judge acquiesced with Andrew. An institution obliged. This was how they handled unmanageable women. They abandoned them in buildings on the brow of the bluffs.

Anna lost her son, her land, and the freedom she had sought when she applied for a homestead. She was the cautionary tale told in the boom town newspapers, the woman sent to Jamestown, never to return.

There at her grave, I could sense the powerlessness she and other women experienced under such circumstances. I knew that what I imagined of Anna's imprisonment was a modern projection of my fear as a woman of being trapped in circumstances I didn't envision, create, or desire. Confinement, in all its meanings. But Anna never had a chance. She was shoved into that asylum sixty-seven days after giving birth, diagnosed as insane, and then forgotten and forced to abandon her baby. She was someone who deserved better. She was worth remembering, worth loving.

Burke County, the place with such lofty aspirations it hired a fancy Minneapolis architectural firm to build its courthouse, saw its land boom go bust in a generation. The Dust Bowl and the Great Depression and World War II all took their toll. My grandfather left as soon as he was old enough to go. He never returned to North Dakota to live there. When he did return to visit, he would have seen a year-by-year, decade-by-decade slide.

Without constant maintenance, the wind strips away the paint and pulls off the shingles and tears the awnings into tatters. Insects gnaw at the boards. Barn swallows and other birds roost inside. Brambles overgrow it all. Without attention, without love, the decay advances faster, unchecked.

Caretaking implies not only a future but also a respect for the preciousness of the present moment. Someone has to love something enough to sweep up the daily dust, to reapply a coat of paint and to patch the leaky roof. Who knows what sort of life Anna would have lived had someone envisioned a future for her outside of the asylum?

A future that wasn't preordained by her own mother's past? A future where she could be loved.

Anna's gravestone lists only her date of death, not the date of her birth. It is the same for most of the people buried there. Few birthdays are listed, and almost none of the graves are marked by epitaphs. In some ways, it makes sense. The people buried in the cemetery weren't born at the state hospital, they didn't live their full lives there. They merely died there. The meager but quiet cemetery was their memorial. It was an eternal respite for those forced to live within an institution of unrelenting cacophony, and within minds that perhaps never quieted. It was nicely kept, as the superintendent said truthfully in his letter to my grandfather.

I sat alone in the cemetery, listening to my breath move in and out. I could hear my heartbeat. A light breeze continued to blow, all around me. The hum of traffic on the highway continued, uninterrupted. Birds sang. The sun, moving farther away by the day, still warmed the earth, even in late October.

Part II

9

A WOMAN BEHIND
EVERY TREE

February 2014

North Dakota crude: *$91.68 per barrel*

"Don't go into Walmart by yourself!" This warning came from everyone I met who'd traveled to North Dakota, and even from people who lived there, men and women both. I knew it was unwise to visit a bar on my own in a place where men outnumbered women, but the warnings about Walmart seemed alarmist. It was just a store—or so I thought.

Walmart, though, became one of the symbols of how ill-prepared western North Dakota was for an oil boom the scale and scope of the one under way. In Williston, the population doubled from thirteen thousand in 2009 to twenty-six thousand in 2015. At the height of the boom in 2013, one-bedroom apartments allegedly cost more in Williston than Manhattan.[1] That was if there were any vacancies.

Walmart in Williston was often the first stop for many new arrivals, and not just for supplies.[2] The company had a generous national policy of allowing RVers to camp overnight in store parking lots. It didn't take long for people swarming to the Bakken oil boom in search of jobs to begin testing the limits of the company's hospitality. "In Williston, those overnights sometimes turn into however long people can get away with being there until they're told to move along, or until they find other, better circumstances," the *Bismarck Tribune* reported in 2012.[3] Not long after, the Williston store banned overnight camping in its parking lot. Managers posted yellow notices on the RVs, ordering people out. "Extended parking in the lot of this store has created safety, noise, litter and property damage problems," the store said in the notices.

Walmart had a dark, albeit random connection to another story that frightened many women in the region. Shortly before the Williston Walmart ordered out the RVs, two men looking for work in the Bakken oil fields in 2012 abducted and killed forty-three-year-old Sherry Arnold, a schoolteacher out jogging in Sidney, Montana, a town near the border of North Dakota.[4] The men abandoned Arnold's body about an hour away, near Williston. Then, they drove to the Williston Walmart to buy a shovel to dig a grave and bury Arnold's body.

It was a terrifying, random crime, and the connection to the retailer was tenuous at best. But Arnold's death and the influx of men without community roots was more than enough for women to fear for their safety at the store. The killers had been there. In that same parking lot.

It wasn't as though I didn't worry about my safety while I was in North Dakota. I did. I was cautious without being timid. But whenever people warned me about Walmart, I was realistic. I told them that statistically, it was far more likely to be involved in a car collision than to be assaulted in a Walmart parking lot. The infrastructure in North Dakota just couldn't keep up with the traffic. There were stories in the news nearly every day about deaths and crashes on roads unprepared for the influx of tanker trucks and fracking equipment and overworked, overly tired oil field drivers. North Dakota had the third-highest rate, per capita, of road fatalities in the country in 2013, I told people. Washington, DC, where I lived, had the lowest.[5]

That said, my first trip to North Dakota, I, too, got a little spooked after I dashed into a Walmart next to my hotel in Minot to pick up some supplies. Even there on the edge of the oil patch, one hundred miles from Williston, there were only a few women shopping in the evening. None were solo but me. I felt the hungry stares of men. Lest anyone think my mere presence was an invitation, I avoided eye contact as I hurriedly picked out produce. Even a married forty-year-old female journalist with messy hair and intimidating posture was seen as just another piece of merchandise.

For many Native women in North Dakota, the threat of violence went well beyond the Walmart parking lot, and it existed long before the Bakken oil boom. Indigenous women are more than twice as likely to be sexually assaulted than other women in the United States. Some federal crime statistics suggest that Native women are murdered at more than ten times the national average. When Native women go missing, their disappearance is much less likely to be investigated, particularly if the perpetrator of the crime isn't a tribal member.[6]

One Native woman's death in 2014 sent shivers down my spine. Robin Fox was thirty-eight, about the same age as me, and had gone missing from a bar in North Dakota the same day I flew home to Washington, DC. The short news account of her death illustrated the heartbreaking lack of concern about such disappearances. "The woman's body was found on the property of a non-tribal member," a brief article in the *Bismarck Tribune* said, without including Fox's name, "so a deputy responded and called the coroner." The sheriff "did not have the woman's name or any other details of the case, since it was deemed a tribal matter."[7]

The article demonstrated the jurisdictional quagmire many Native women and their families face if they are victims of crime. Often, tribes have little criminal jurisdiction over non-Natives who commit crimes on tribal land, particularly domestic and dating violence. Activists continue to work to address such gaps in federal and state law, and led by Indigenous women, had just successfully pressed in 2013 for the reauthorization of the Violence Against Women Act. The bill, signed into law by President Barack Obama, had language allowing some tribal courts to prosecute non-Natives in domestic violence cases.[8]

Robin Fox had a nursing degree and was the director of preschool.[9] She was beloved by friends and family. "Robin loved to dance and had a smile that could light up a room," her family wrote in her obituary. "She excelled at everything she did. She had many friends and was good at sharing a joke. She had the kind of laugh that when you heard it, it made you laugh. Robin loved her family and cherished spending time with them." Her death may not have been directly tied to the oil boom, but the lack of investigative concern reflected the continuous peril for Native women in a place shaped for centuries by the violence of men and extractive industries.[10]

The oil boom brought wealth and its accompanying problems to the Mandan, Hidatsa, and Arikara tribes, known as the Three Affiliated Tribes or MHA Nation. By 2014, the tribal nation had received more than $1 billion in oil tax revenue from the boom. Oil wells on the tribe's Fort Berthold reservation in west-central North Dakota were responsible for roughly a third of the state's oil output. The Federal Reserve Bank of Minneapolis noted that if the tribal government were a state, the MHA Nation would rank the tenth biggest oil producer in the country.[11]

Just like in the rest of North Dakota, who ended up with mineral rights on tribal land was often pure luck and happenstance. The new wealth often only exacerbated existing economic disparities and social problems—on the reservation and beyond. The boom was accompanied by a crime wave in oil-producing counties of North Dakota. One analysis by researchers at the University of North Dakota and the

University of Regina in Canada in 2014 found that violent crime in oil boom counties increased 18.5 percent between 2006 and 2012.[12]

People in Williston were afraid, even if crime never touched their daily lives. The day before Halloween in 2012, the Williston newspaper printed the names of all the registered sex offenders living in the city. It was accompanied by a fear-stoking headline: "Number of sex offenders on the rise."[13] The subtext wasn't hard to read: the increase was caused by outsiders flooding the state in the wake of the recession, in search of jobs.

The Overnighters, a documentary by filmmaker Jesse Moss, captured some of the anxiety of the time. Released in 2014, the film featured Jay Reinke, a pastor who opened the doors of Concordia Lutheran Church in Williston to new oil field arrivals who couldn't find anywhere to live, including some sex offenders. Reinke's ministry split the congregation. "I didn't feel safe coming into the building," one parishioner admitted on camera. Eventually, Reinke stepped down as pastor.[14]

The story of itinerant men and their behavior during the boom seemed so obvious it didn't interest me. It was ground well-trodden by other journalists, a story as old in America as the California Gold Rush, if not even older. Plenty of other people were writing about the influx of strippers in Williston, or life in "man camps," the temporary housing used to house oil field workers who arrived in the state in the earliest days of the boom. Even the *New York Times* weighed in with a piece about nightlife in Williston: "Natasha, 31, an escort and stripper from Las Vegas, is currently on her second stint here after hearing how much money strippers made in Williston," the paper wrote breathlessly in 2012. "'We make a lot of money because there's a lot of lonely guys,' she said."[15]

There was even a joke in treeless North Dakota, about there being a woman behind every tree. In Montana, a friend told me, the joke was that in North Dakota, there was a *pretty* woman behind every tree.

I checked in with an editor at *E&E News*, which paid me to write freelance pieces from North Dakota. I wanted to be writing articles about the effect of North Dakota's oil boom on the lives of women as well as on schools, roads, and the environment, particularly climate change. But should I, too, be writing about man camps? How many stories could you do about places that smelled of sweaty socks, energy drinks, and farts? Hadn't men controlled the narrative here for long enough?

"Ugh, man camps," my editor said, rolling her eyes. "We've written so much about man camps."

I felt vindicated. Shortly after my first trip to North Dakota, I was a finalist for a prestigious journalism grant that paid $40,000. I proposed writing articles about the effect North Dakota's oil boom was having on women. In my final interview for the grant, one of the judges walked me out of the room, shook my hand vigorously, and told me I was doing important work that should be funded: "I really want you to get this," he said, looking me directly in the eye.

"Nailed it!" I texted my husband, as I walked to the Metro.

But I hadn't. I was one of ten finalists for four slots, but other journalists got the grant. One went to a team writing stories about oil pipeline safety, and another went to a duo writing about the complexities of reintroducing bison to, of all places, the Great Plains. North Dakota stories were hot that year, and it stung me that the judges thought other people had better ideas than I did. I dusted myself off and made plans to return for more research, with a side trip to Colorado. I was free to pursue the story of the woman who brought me to North Dakota: Anna.

———

Dave Kornder stood with his back to a panoramic view of the Colorado Rockies, visible from the windows in the 16th floor conference room of the downtown Denver offices of Cornerstone Natural Resources.[16] I was seated at a long conference table opposite him. Atop it sat a narrow plexiglass case, about five inches wide, two inches tall, and five feet long. It was filled with drilling core samples from the various shale rock formations in North Dakota that yielded fracked oil. Samples from the Three Forks formations ranged from a dark gray to a tannish cream-colored rock. The Middle Bakken Reservoir rock was a sandy tan that would have looked at home on a modern kitchen countertop.

I was in Kornder's downtown Denver office to make a big ask: I wanted to follow along with Cornerstone crews if they drilled for oil on the land my great-grandmother once homesteaded. At the time, Cornerstone Natural Resources held a majority of the leases in Burke County, the place where Anna claimed her homestead. Those holdings included the lease my mother, my uncle, and their two half-sisters signed in 2009.

My time was limited. I was in Denver for barely twenty-four hours, just enough time to interview Kornder and two scientists with the U.S. Geological Survey. Then I was returning yet again to North Dakota for more research on Anna, this time in the winter. I also had two free-lance assignments that would help me pay for all the travel.

I pressed my case for the next hour, trying to pry more information from Kornder about his company and its operations. I mentioned that, like him, my parents and my sister and an aunt and an uncle all attended Montana State University. I appealed to his vanity, flattering him for his management and unconventional business approach. I reminded him that his company's story was a very American yarn about trying to make it on the prairie. It was no different than that of my pioneering great-grandparents, I told him—even though I didn't let him know that I had lost nearly all faith in the myths of the West.

I did everything but what I suspect he really wanted, which was for me to walk out the door and leave him alone to make money.

This was, I was beginning to understand, a common attitude in the North Dakota oil fields, especially among the men who dominated its business and political worlds. There was little room for criticism of the boom, sometimes even in the face of tragic events.

Just two months before our meeting in Colorado, a train carrying crude oil collided with a derailed train carrying grain, near the town of Casselton.[17] No one died, but the fire burned for more than twenty-four hours. The crash came six months after a runaway oil train jumped the tracks and caught fire in Lac Mégantic, Quebec. The resulting inferno killed forty-seven people and destroyed much of the downtown.[18]

In the days following the North Dakota train collision, an energy lobbyist and prominent Republican in the state, Robert Harms, told a reporter with Reuters that a "moderated approach" was needed. "I think it's a good wake up call for all of us, both local and state officials, as well as the people with the oil and gas industry and the transportation industry. Even people within the oil and gas industry that I've talked to feel that sometimes we're just going too fast and too hard."[19]

The most prominent person with the oil and gas industry in North Dakota was swift to suggest otherwise. "We shouldn't overreact," Continental's Hamm told the *Dickinson Press*, one of the newspapers in western North Dakota. "We're doing something that's not only the best thing for North Dakota and for that area up there, but also for our entire nation. The world has been changed by the fact that we can produce energy of this quantity in America today."[20]

In the face of industry blowback, Robert Harms walked back his comments. It all played out online. "We aren't faced with one of two choices: either the status quo or shut down the oil industry," he wrote on a widely read conservative news site. "The event in Casselton should give us pause to have thoughtful discussions."[21]

But the prevailing view in North Dakota was that "thoughtful discussions" were a threat to the policies that allowed oil companies to drill fast, extract what they could, and then leave, consequences be damned.[22] The only thing that mattered to oil companies was walking away from North Dakota with as much money as possible.[23]

Even so, some people were worried about the impact of the oil boom. The previous fall, as I drove across North Dakota after visiting Anna's grave, I had listened to breaking news reports about a ruptured pipeline in Tioga that spilled eight hundred forty thousand gallons of oil in yet another wheat field. The oil spill, which cost $93 million and multiple years to clean up, was not far from the original well that Bill Shemorry photographed in 1951.[24]

And then there were the radioactive filters that kept turning up all over western North Dakota.[25] Known as socks, the tubular filters collected naturally occurring contaminants from saltwater pumped out of the earth as part of the fracking process. Because of their radioactivity, they had to be disposed of in special hazardous waste landfills outside of North Dakota. In early 2014, thousands of filters were found discarded not far from Anna's homestead in an abandoned gas station in the tiny town of Noonan near the Canadian border, population 120.[26] A Montana man who was paid to dispose of the radioactive filters in a landfill instead shoved them inside more than 200 industrial-sized trash bags and then discarded them in the building. The man responsible for the mess, James Ward, was for a time the federal Environmental Protection Agency's most wanted fugitive. (Ward was not sentenced for the crime until 2020.)[27]

One landfill operator, Rick Schreiber, installed a radiation detector to find errant socks. In 2013, Schreiber was uncovering as many as ninety socks per month in the landfill he ran for McKenzie County in western North Dakota. Schreiber complained about the dumping to local, state, and federal environmental regulators.

"Somebody has to take a stand," Schreiber told a reporter with *High Country News.* "I'm not here to make friends with the oil companies, I'm here to do my job and...make sure my legacy here in 20 years...isn't destruction of the environment and a Superfund project."[28]

Those like Schreiber who were worried about the impacts were outspoken in a Midwestern culture with little social tolerance for dissent.[29]

Valerie Naylor, the former superintendent of Theodore Roosevelt National Park, told one newspaper at the height of the oil boom that she feared taking time off, just in case oil companies began drilling within view of the park while she was away from the office.[30] Wayne Stenehjem, the state's Republican attorney general at the time, proposed to the North Dakota Industrial Commission that it establish extra regulatory steps if oil companies wanted to drill in or around eighteen so-called extraordinary places, including state and national parks.[31]

Stenehjem's proposal[32] was doomed, thanks mostly to industry influence, including the North Dakota Farm Bureau and Continental's Hamm. "Sadly, the policies have stolen a page right out of the environmental movement's playbook," Hamm wrote in a letter to the Industrial Commission in early 2014.[33]

The Bakken, Hamm warned in his letter, "is not the only attractive play in America, and other states would be all too happy to have an investment supporting job growth and energy independence with their borders." Hamm's threat to leave North Dakota wasn't even veiled; it was right there in the public record for anyone to read. He was warning the state's leaders that Continental could take or leave North Dakota, especially if conditions were more welcoming elsewhere.

The state adopted a watered-down version of Stenehjem's bid to save extraordinary places. One of the few environmental organizations in North Dakota, the Badlands Conservation Alliance, took a surprisingly chipper view of the proposal's failure in their year-end summary of their work advocating for the state's wild places: "It was the first time a high-ranking state official actually said there even were places worth protecting from development," the group wrote.[34]

That day in Denver, Kornder grudgingly walked me through the basics of his company's operations and business strategy in North Dakota. He pointed to the outer edge of a cluster of tiny yellow dots on a map of North Dakota on the wall of his conference room. They represented all the wells being drilled in the Bakken. When Kornder took over the company in 2008, he decided to target the outer cluster of dots. The company already held a number of leases in the area, in a separate, much younger geological formation known as the Nesson. Company executives thought they would see whether pursuing the outer edges of the Bakken could be a profitable approach.

Away from the center of activity, Cornerstone found there was less competition for leases, so it cost less per acre to drill. Major oil companies also had less interest, freeing the fringes for independent operators. He told me that the oil being fracked out of the wells on the edge of the Bakken had higher water content, making it more expensive to process because the company had to separate out the water and dispose of it. But those costs were offset by something other producers were struggling with in North Dakota: easy access to natural gas pipelines. Unlike many other producers in North Dakota, Kornder told me, Cornerstone only had to flare two of its existing forty-four wells at the time. That meant natural gas was also an additional revenue stream for the company.

Kornder asked me to send him several of my previous articles about North Dakota. I thought I was sharing unbiased reporting; it seems he saw a journalist out to unearth bad news in everything I wrote. I even gave him the name of a Republican energy lobbyist I knew in Washington, DC, who could vouch for my evenhandedness, but Kornder was unpersuadable.

He made it clear: he told me he didn't like journalists and he saw no benefit in my request to tag along every step of the way as his company drilled on the land my family once owned.

I knew his type, and it was nearly impossible to convince men like him to talk. They believed you were either with them or against them. There was no middle ground, no nuance in their world, especially when someone without power or influence stood in the way of what was more widely seen as progress. This very same attitude was on display in Hamm's letter to the Industrial Commission.

Mostly, though, I don't think Kornder wanted a journalist tagging along as his company fracked its next well in Burke County. He didn't want *any* press, I suspected, let alone the risk of bad press. If all went according to plan, Kornder and his investors would become even richer from gambling on long prospects in the ghost towns and wheat fields on the edge of the boom.

He told me flat out that all I'd probably get from him was an hour in a conference room on a Thursday afternoon in late February. Nonetheless, I sent him a thank you note from North Dakota—a postcard of a pumpjack silhouetted by the sunset. But he still said no.

All that "no" was dispiriting enough when I was a reporter at a newspaper company with declining fortunes. Now, out on my own, it felt even worse to be denied access by a wealthy man whose interactions with me suggested he thought his company and the industry itself should be above all scrutiny.

Yet it was doubtful Kornder had much to hide. He wasn't in Hamm's league, and neither was his company. What was far more likely was that I didn't matter to him. Denying me access to the story I wanted to tell about my family made me feel small, but it wasn't a novel experience—all women have stories of men exercising their power to make us feel inconsequential. The meeting on the 16th floor of that downtown Denver office building was yet another reminder that life in the oil patch was very much a man's world.

As I walked out of his office, it was a sunny, late-winter day in Denver, warm enough to sit outside and eat a sandwich downtown. I reminded myself that my favorite and most significant stories were always the ones I uncovered when I didn't have access to power brokers or important people. When you aren't an insider, you have to be creative to find good stories. You have to look sideways, into the nooks and crannies that other people neglect. I didn't need permission from a rich oilman in Denver.

––––––

Two women ended up salvaging my trip to Colorado: Stephanie Gaswirth and Kristen Marra.[35] The two scientists, with the U.S. Geological Survey, had offices at the sprawling federal government campus in Lakewood, a suburb of Denver. The two were the authors of a 2013 government report that estimated the amount of oil in the Bakken and Three Forks geologic formations in North Dakota and Montana.

In 2008, the U.S. Geological Survey issued a dryly worded assessment of the Bakken that, for those paying attention, confirmed what many oil companies already knew: there was a lot of oil in North Dakota.[36] The Bakken, the government said in 2008, held an estimated 3 to 4.5 billion barrels of oil, "larger than all other current USGS oil assessments of the lower forty-eight states and...the largest 'continuous' oil accumulation ever assessed by the USGS." The USGS report foreshadowed the coming boom. It was a big deal.

But several years into the boom, oil companies and officials in North

Dakota pushed for the USGS to update its assessment. Horizontal drilling technology was changing fast, and it was getting easier, faster, and more predictable to access oil in North Dakota. Because there were more wells, oil companies and geologists had more information about what was under the earth. They knew the 2008 assessment was already outdated.

The oil companies and the state of North Dakota didn't just want a big number, they needed it. A new assessment from the USGS, the only unbiased provider of publicly available estimates, would help oil companies persuade investors and shareholders to back new prospects. Fracking is a cash-intensive business that requires expensive equipment and labor outlays before oil or gas ever begins flowing. Energy companies also need cash to pay leaseholders, even if they never drill on the land. Low interest rates after the Great Recession made borrowing cheaper. For investors with cash, fracking was one of the few places to put their money that offered growth during the recession.[37] A big number on the new USGS assessment meant some people would get even richer.

The Obama administration agreed to revise the 2008 report. Gaswirth and Marra got to work. All the new wells in North Dakota and Montana gave the scientists much more information to complete a new assessment. Their update, released in April 2013,[38] doubled the amount of technically recoverable oil to 7.4 billion barrels in the Bakken and Three Forks geological formations. It was enough oil to fuel the U.S. for a year.

But no one in power in North Dakota thought the USGS estimate was big enough, and many were disappointed with the assessment. One critic called it "a milemarker that's well behind you in the rear-view mirror."[39]

Many, including Continental's executives, were literally banking on it being much, much bigger. Hamm told the *Oklahoman* newspaper it was too low—his own geologists estimated the Bakken held twenty-four billion barrels. "The USGS generally is very conservative in their estimates," he said.[40]

Gaswirth and Marra told me about a curt phone briefing they had with elected officials in North Dakota before the Interior Department made the USGS report public. Among them were U.S. Senator of North Dakota John Hoeven, a Republican who had been governor during the start of the boom.

"Is that it?" asked Hoeven. He, along with Hamm, was among those who urged the federal government to reassess the amount of oil. "Okay," Hoeven said, and then hung up the phone.

The call characterized the reaction among those with power. After the assessment, Gaswirth told me, Continental Resources stopped providing USGS with any additional geological information from their North Dakota drilling operations. "So they were very helpful until the number came out and it was much lower than I think what they wanted," Gaswirth said. "Then we stopped hearing from them, and we're no longer talking to them."

I considered myself lucky to be talking to the two geologists. They didn't have an agenda. They were scientists who based their findings on what the data told them, not the politics or financial markets.

Sitting at her computer, Gaswirth zeroed in on Anna's homestead. She used subscription mapping software that showed the geology and yield of other wells nearby. There weren't many other wells. Most were older, more conventional oil wells that tapped straight down into formations with reservoirs.

She gave me a quick geology lesson. Oil in the tight, shale rock formations of the Bakken emerged from organic matter that was deposited on the bottom of very quiet, very deep water environments roughly 350 million years ago, in the Devonian period. The oil in the Bakken is especially attractive because it's easy to process, Gaswirth said. "It's a very high quality oil. People like producing this type of oil."

She pointed to her screen.

"You're on the western part? So in here somewhere?"

I nodded.

"That's interesting," she said.

I cringed. "Interesting" was often a softener for news you may not want to hear.

"That's better than being in the east," Marra assured me.

"It's closer to where the majority of the oil wells are," Gaswirth said.

Fracking made feasible what seemed impossible to many oil producers in North Dakota only a few years earlier, when geologists still used more conventional oil drilling techniques. As the oil boom matured, geologists figured out how to get the most out of the earth with horizontal drilling practices. The more that oil companies fracked in North Dakota, the more efficient they were.

But what these two women were telling me was that Anna's land was unlikely to yield riches. They knew that oil companies would seek out the easiest prospects first, not the watery margins, like Anna's former homestead. They thought it was unlikely anyone would ever drill on Anna's land, not when so many better prospects abounded in North Dakota that wouldn't be as costly or have such long odds of success.

I felt an unexpected wave of disappointment. I had been rooting for Anna's homestead to yield a windfall, I realized. The toxic myths of manifest destiny shaped me, too. I wanted to believe I was above the tug of the whispers, but I had heard them my whole life, too. They were nearly impossible to ignore, especially as an American whose ancestors strode ever westward chasing their dreams. The pursuit of happiness was written into our very Declaration of Independence. It was what set Americans apart, we were all told.

The pang of the letdown lingered with me as I rode the shuttle bus from the rental car lot to the Denver airport, and then boarded my flight to North Dakota. On the flight, I kept mulling over what I learned in Denver. I was thankful for the professional assessment of the prospects of the mineral rights my mother inherited. It was a fun interview with lively conversation with other women who loved what they were doing. Gaswirth and Marra were as intrigued by my research as I was with theirs. I could see how curiosity shaped their careers, just like it shaped mine. In ninety minutes in Gaswirth's office, I felt like I learned more about geology than I did in a whole semester in college. I walked away with new knowledge about the world we live in.

Other than the flight attendants, I was one of only a few women on the plane from Denver to Minot. Most of the other passengers looked like energy executives, or maybe engineers or geologists. They wore fleece jackets with company logos over tucked-in button-down shirts freshly pressed with the sort of creases that come from the dry cleaners. Their nails were tidy, and they had sharp haircuts and clean-shaven faces.

Those guys, the ones making all the money and decisions, they didn't have to overnight in a parking lot or the basement of a Lutheran church while they looked for a job in the oil fields. Someone back at the home office made reservations for them. Someone made sure a Suburban awaited the executives at the dumpy airport, to whisk them away to wherever it was they were going.

The executives were still on my mind as I checked into a hotel in Minot, using credit card rewards miles I scrounged together to afford the trip. None of those men had to look over their shoulder just in case someone was following them from their car to their hotel room. It probably never occurred to them to befriend hotel clerks like I did, just so there was a witness who remembered my name and face and last known whereabouts. They didn't think twice about the consequences of going to bars or restaurants alone. They certainly didn't worry about the Walmart parking lot.

This was still a man's world, just like it had been for Anna.

10

LONG UNDERWEAR

March 2014

North Dakota crude: *$91.17 per barrel*

IT WAS TWO DEGREES BELOW ZERO OUTSIDE. I PULLED MY BLACK CASH-
mere turtleneck over my long underwear and my jeans. It was the
same sweater I had worn for comfort in the cold days after my mother's death, and the armpits were beginning to go gray and thin. But I
couldn't bear to toss it, not yet. It was still my warmest sweater, the one
I reached for when I needed to feel coziest. I donned a pink hat, gloves,
and my black, fleece-lined North Face ski jacket. There's no such thing
as bad weather, I told myself as I walked outside to my rental car, just
bad gear.

I was on my way to one of the weirdest facilities in North Dakota's
energy portfolio: the Great Plains Synfuels Plant near Beulah, a town
eighty miles to the northwest of Bismarck. The plant used low-grade
lignite coal mined in the surrounding fields and converted it into synthetic natural gas, other fuels, and fertilizers. It boasted of being the
only commercial-scale coal gasification plant in the United States to
manufacture natural gas.

The cycle began with lignite coal unearthed from surface mines nearby. The coal was burned in a power plant adjacent to the gasification plant. The electricity generated at the power plant fueled the manufacture of synthetic natural gas, which was extracted from even more of the lignite coal.

The gasification plant captured carbon dioxide, the inevitable byproduct of making synthetic natural gas from a fossil fuel dug from the earth. Then, the captured carbon dioxide traveled 205 miles north in a pipeline to the Weyburn and Midale oil fields in Canada. There, the carbon dioxide produced in North Dakota was sequestered beneath the earth in Saskatchewan, but not until it was used to force more oil from aging oil wells drilled in the mid-1950s. The oil unearthed from the Canadian oil fields was turned into gasoline and other petroleum products, releasing even more carbon dioxide into the atmosphere once it was consumed.

The plant was a $2.1 billion relic of another energy age, a costly Rube Goldberg machine conceived of during the energy crisis of the late 1970s as a way of producing natural gas in the event of another embargo. It seemed especially baffling that the plant manufactured natural gas, given how much methane was already being flared from the Bakken oil fields in North Dakota. And yet the gasification plant still existed a generation after its construction, running a big, dirty coal-fueled cycle.

As I drove up to the plant, clouds of steam billowed from the smokestacks. I was there to write about Gina McCarthy, the administrator of the federal Environmental Protection Agency.[1] She was on a multistate listening tour to drum up support for the Clean Power Plan, the Obama administration's plan to address climate change by reining in greenhouse gas emissions.

Few people in North Dakota supported the president's new rules for power plants. Yet McCarthy, a blunt-talking public health official with a distinctive Massachusetts accent, seemed undeterred as she entertained questions inside a plant that not only burned coal, but actually also manufactured planet-warming greenhouse gases. She matter-of-factly told the workers and public officials that no matter what she did, she was going to be sued by environmentalists and utilities both.

At the time of McCarthy's visit, about 75 percent of the electricity generated in the state derived from eight coal-fired power plants. People at the gasification plant were worried about keeping the state's coal plants open if new carbon emission rules took effect. North Dakota's coal production was small compared to Wyoming or West Virginia, but in 2014 coal was the main source of fuel for the power plants that heated most homes and businesses in the state.

The lignite coal mine, the power plant, and the gasification plant made up the bulk of the tax base in the rural county. If the plants closed, jobs would leave and schools would atrophy, the energy executives, union officials, state regulators and elected officials told McCarthy. Just step outside, one man at the plant told McCarthy, and you'll see why a stable source of electricity matters in such a cold place.

"The last thing we want is to have this rule impact the ability of the economy to grow in any state, never mind nationally," she told them.

I was curious to see whether McCarthy's reception would differ from Sally Jewell's tour of the oil facilities with Hamm, just six months earlier. Both cabinet secretaries showed up in the state to tell North Dakota it had to find a way to scale back greenhouse gas emissions. Both women represented a future the state wanted not only to deny but also to defy. People were "North Dakota nice" to McCarthy, a phrase that reflected the passive-aggressive nature of public discourse in the state: pleasant on the outside, seething on the inside.

When it was time for questions, though, someone in the audience confronted McCarthy directly: "Is there a war on coal?"

"No," McCarthy told them.

No one believed her, including McCarthy's host, Heidi Heitkamp, then a Democratic senator who represented North Dakota in Congress. Heitkamp, who before her election sat on the board of directors of the gasification plant, referred to the Obama administration's climate policy as one that was "hell-no-to-coal."

But the truth was unavoidable. A war on coal wasn't even necessary. Fracked natural gas was plentiful and cheap, and despite valid worries among environmentalists, the Obama administration considered it a less harmful source of emissions than coal. Natural gas already was replacing coal-fired plants around the nation. Even in North Dakota, wind turbines were sprouting everywhere across the windy prairies. Over the next seven years, despite the recalcitrance in North Dakota coal country, wind power became nearly one-third of the state's energy mix.

———

The temperature dropped steadily as I drove toward Williston. So did my gas gauge. I'd started out that day in Minot, and then I drove south to Bismarck to see McCarthy speak at a community college. Then, I drove west to the event at the plant. I had been so busy following McCarthy, it didn't occur to me I'd already driven two hundred miles that day. By the time I realized I was low on fuel, I was at least fifty miles from the nearest gas station.

As I drove, I grew more and more alarmed. I'd chosen a scenic route so I could take photos, which seemed a foolish decision when it was too cold to exit my vehicle. Few other cars were on the road. My cell service was poor. I searched Google maps for a gas station, but nothing came up. I was in the middle of nowhere, it was getting dark, and it was -4 degrees outside. My mind mulled over the most cata-strophic scenarios. It didn't help that I'd recently read *The Children's Blizzard*, a book by David Laskin about the young people who froze to death in the freak 1888 storm that swept the Great Plains. The book ended with a hauntingly beautiful line: "After that day, the sky never looked the same."[2]

I coasted into Watford City on fumes, seventeen miles left until empty, according to the gas gauge. At the station, I filled the tank and bought snacks. As I drove another fifty miles toward Williston, I munched on caramel-flavored Bugles and considered what I would have done if I'd run out of gas on a rural road in subzero temperatures. It turned out fine, but it was the least cautious I'd been in all my time in North Dakota so far. Driving was when I faced the greatest risk to my safety, I knew, and I promised myself not to allow the gas gauge to dip below a quarter tank. It became my own grim joke whenever I filled up: "After that day, the gas gauge never looked the same."

The next morning, it was twenty-two degrees below zero when I left the hotel and drove toward the offices of Empire Oil Company on the edge of downtown Williston. The blue two-story building looked like an apartment complex from the outside. Inside, it was an ordinary office, like a title company or bank. Empire Oil didn't actually drill for oil, it just took care of leasing paperwork, including the lease my parents signed in 2009.

There, I met with Ian Vestal, a landman who had authorized thou-sands of leases and who oversaw Empire's activity in Burke County.[3] Vestal, whose family owned a trucking company in Williston, told me he had studied journalism in college. He was eager to help in any way he could, a friendliness that came as a relief after my visit with the oil executive in Denver.

Vestal pulled my family's file. Empire's landman began in 2009 with a records search not unlike my own at the Burke County courthouse. He started with Anna's patent, the document the federal government issued in 1912 when her husband, Andrew, sought title to the land. Then, the landman traced each and every transaction on the property, going forward in time. The work included looking at second mortgages, liens for unpaid taxes, foreclosures, and previous leases. It was, I saw, far more thorough than my own search in Bowbells.

When my grandfather sold the land in 1973, he kept 40 percent of the mineral rights. Empire's search showed that the subsequent owners of the property lost both the land and the remaining mineral rights in the 1990s. The land went up for sale in a sheriff's foreclosure auction, he told me, and a local farmer ended up with it.

At the time, the farmer who bought the land didn't snatch up the remaining 60 percent of the mineral rights, though. Somehow, those rights ended up with an oil company based in Maryland. The Maryland company, along with my mother's heirs, my uncle, and two half-aunts, would reap any royalties from drilling on Anna's original 160 acres.

If anyone ever drilled, that is. Just like the scientists in Colorado, Vestal was skeptical it would ever happen. Oil companies were signing thousands of leases on land where they had no intention of drilling, he told me. They were doing it to throw off competitors, to hedge their bets and to cover their bases. Companies like Hamm's Continental borrowed billions of dollars in complicated bond deals to pay for leases.

"They can't get to it all," Vestal said. "Nine out of ten leases will expire. Out of 10,000 leases, they're drilling on 5 percent. They'll spend $20,000 to save $5,000."

This was the dirty secret of the oil boom. Production investment was rewarded, not profitability. Investors put their money into oil companies for their potential riches, not their current cash flow. That meant oil companies had to keep throwing a lot of cash at leases and potential oil wells to show investors their growth potential. Thanks to the Great Recession, interest rates were low, and it was cheap to obtain money to write checks to people like my mother. For certain types of investors with ready cash, fracking at the time looked like an attractive sector with unlimited growth.

But it wasn't financially sustainable, not unless oil prices stayed high long term. Oil trades on a volatile global market, subject to geopolitical whims far from the prairies of North Dakota. It was inevitable that prices would once again go down.

Vestal's friendly recap left me with the same contradictory pang of disappointment I experienced after visiting the USGS offices in Denver. From what everyone was telling me, the lease payments might continue for a while, but no one in my family would ever get rich off the mineral rights Anna left behind.

As Anna's descendant and as a storyteller, I ached for her life to have some meaning. I longed for a narrative arc that fit tidily into what's known as the "and-but-therefore" formula beloved by Hollywood. I could even imagine the dramatic language of a movie trailer, voiced over soaring strings and prairie vistas: *In a windswept land, a pioneering woman homesteaded the untamed prairie, in search of her own slice of the American Dream. But life took a tragic turn, and she died unknown and unloved in an insane asylum. Her land and her legacy sat forlorn, for years, forgotten until the oil boom came along. Out of tragedy emerged triumph, redemption, and fortune—for generations to come.*

Like so many myths of the West, I knew it was an exaggerated account of what happened, weighed down by wishful thinking and my own attachment to the story lines of capitalism. Yet the meaning provided by capitalism was what I—and most Americans—knew best.

The movie trailer of my daydreams failed to acknowledge the

other grim consequences of the pursuit of wealth in North Dakota. All those files in Vestal's office represented piecemeal decisions by individuals and families who owned a patchwork of mineral rights. They made their decisions 160 acres of homesteaded land at a time. It began with wresting the land away from Indigenous people, and ended with all that flared gas speeding up climate change. Along the way, millions of acres of prairie got plowed under, changing the entire ecosystem of the Great Plains. Economies went boom, and then they busted. Men got injured. Native women went missing. Women like Anna were abandoned in asylums, left to die there.

The decisions we made as individuals in pursuit of wealth had collective consequences on the planet but also the wellbeing of our communities. Only a few people grew rich, and at what cost? I wanted to believe an alternative story line was possible.

———

Wary of slipping, I penguin waddled carefully along the frozen sidewalk toward the entrance of a 231,000-square-foot municipal building with a biblical name: the ARC, or Area Recreation Center.[4]

Construction was almost complete on the $76 million behemoth, dedicated entirely to fun. (The No. 1 rule of the ARC is to have fun, literally. It's painted on the mural that visitors see when they walk in.) I was there for a pre-opening tour with Darin Krueger, the executive director who pushed for the community to build the place. Krueger was so busy before the opening the only time he had to meet was on a Saturday. "I get about ten of these requests a week to interview for some magazine," he told me, which I knew was a mild exaggeration.[5]

Inside, we walked by baseball batting cages and a golf simulator and, perhaps most importantly for kids in a cold climate, a three-story indoor water slide. Krueger showed me the Olympic-sized

competition pool and a splash fountain made to look like an oil derrick. The amenities were impressive: four basketball courts, an indoor
playing field, four indoor tennis courts, an indoor track, and a playground. The ARC was so big Krueger had to buy Segways for his staff
to get around.

"You'll have to frack for natural gas just to heat the place," I joked
with Krueger, only half-kidding. The town of Utqiagvik on the North
Slope of Alaska did just that. I had been there in 2008 (when it was still
known as Barrow) and had seen how people threw open their doors to
the bitter cold, it was so warm inside from cheap natural gas heat.

At the time, Williston was at the heart of the fastest-growing county
in the United States. At least thirteen thousand new people arrived
between 2010 and my visit to the ARC in early 2014, doubling the
town's population. All those new people made for crowded schools,
roads, playing fields, and parks.

Like the parking lot of the Williston Walmart and the Lutheran
church featured in *The Overnighters*, the old recreation center drew tran-
sient workers. The newcomers were buying one-month passes so they
could shower in the rec center locker room and use the free wireless
internet. But they were terrible houseguests. They left beard trimmings
in the sinks and they lingered in the public areas, which alarmed parents with young children. Someone left a pile of human feces in a
corner, a discovery that made the local newspaper and served as yet
another symbol of the strain on the town's resources.

The booming towns of the North Dakota oil patch were desperate
to do something to meet the needs of all the new people and those who'd
lived there for generations, through multiple boom and bust cycles.

It reminded me of a scene in that documentary about the 1951
oil boom, *American Frontier*, the one that the American Petroleum
Institute made to persuade farmers in North Dakota to sign leases.

In it, community leaders form a Citizens Petroleum Committee. Nils Halverson, the main character, has already signed a lease with an oil company on his father's wheat farm, but he's skeptical about the effects of a boom on the Williston Basin. He asks the existential question that no one else on the committee will: "What happens when they find it? A wild scramble? Boom and bust? What happens? I don't know. I'm just asking. I want to know."

They invite the oil executives to a public meeting. Nils holds their feet to the fire: "I was just wondering: How long is all this going to last?"

The oil executive, a fat cat–type with an oil patch drawl, sizes him up: "I guess what's on your mind is the old story of boom and bust."[6]

The old story, indeed. As oil workers flooded western North Dakota, Williston and other towns still grappled with some of the same issues, especially housing. As the Bakken oil play took off, people in Williston had few options to address rapid population growth. Most money from oil and gas taxes went to the state. Williston and other cities in western North Dakota wanted to see state funding formulas recalibrated so that more of the money came back to the places hit hardest by the boom, instead of mostly into the state's general coffer.

Some communities didn't think they ought to be saddled with more local debt, especially at a time when state coffers were brimming over with billions in oil tax collections. In some small North Dakota communities, towns had maxed out their borrowing capacity with pre-boom projects. It was particularly bad in Williston, where the town was still paying off debt from borrowing too much money to support infrastructure during another oil boom and bust cycle in the late 1970s and early 1980s. The bust in the 1980s was big, and it made Williston wary of sizable investments. Towns in western North Dakota needed the state either to change its tax formulas, or to help them pay directly for more of the urgently needed projects.

So, the people of Williston had to be creative if they wanted nice parks or a fancy recreation center. In Williston, the parks board saw an opening—many North Dakota communities have separate park and recreation boards, which have taxing authority and elected representatives. If they couldn't get the money from the state to build the ARC, at least they could ask local voters to approve a penny sales tax to pay for their monument to fun.

The sales tax failed by thirty-one votes the first time voters considered it, mostly because the parks district vastly underestimated the money it would collect, Krueger told me. They thought they would bring in $1.8 million per year from sales taxes. It turned out that sales tax receipts were so high from the influx of people to Williston buying stuff, that the tax would have collected more than $1 million each month. It was an excessive amount of money for parks. The city's parks boosters adjusted the tax rate, then put it up for a vote again. Voters approved the tax and a new recreation center.

Taxes and bond issues might seem boring, but as a journalist who covered local government for newspapers for most of my career, I knew how crucial they were to communities. I understood how big public projects got built and how much they mattered. I'd written plenty about bond issues for sewer projects and parks and jails and roads. I wrote so much about the local effects of the 2009 federal stimulus package that I once had a sticky note above my computer monitor to remind me of the exact amount for my stories: $787 billion.

In my experience, few small towns could afford anything as extravagant as the ARC. Nor could big cities, for that matter. Nonetheless, Williston built its ARC as though the population might double again, as though people might, two-by-two, arrive in a cloud of dust and diesel fumes and repopulate the northern Great Plains.

This recreation center was bigger than its palatial size. It was a symbol, a very large one, of what happens to a place when it changes fast, and how the people who live there reckon with the transformation. It was an insistent physical assertion on the part of a dusty oil patch town that its inhabitants deserved something nice after their population doubled in just four years. It was both aspirational and foolish. And so, of course, I was drawn to it.

No one openly criticized the giant facility. But there was talk of its extravagance, especially outside of western North Dakota. I talked to Vicky Steiner, a Republican state representative from Dickinson, another oil patch town 130 miles south of Williston. She was also the executive director of the North Dakota Association of Oil and Gas Producing Counties, and she had heard the complaints. If Williston had the money for such a big rec center, why was it begging the state legislature to rethink the funding formulas?

"Williston has the right to be a beautiful city, just like every other city in the state," Steiner told me.

It's true, the people of Williston deserved a shiny new place for fun. Every community does. Yet every time I was in North Dakota, I photographed the evidence of discarded dreams: the abandoned churches and community halls on the prairie, and the old barns and decrepit homesteads left to collapse in on themselves. There was even an abandoned Cold War–era radar station south of the Canadian border not far from Anna's place.

I'd heard from a state lawmaker that North Dakota's tourism department didn't like to publish images of ruined barns or decrepit homesteads. They didn't want outsiders to think North Dakota was in a state of decay, even if relics of past boom and bust cycles litter the state. Regardless, several professionally led photography safaris are happy in the summer to escort out-of-state visitors on tours to photograph ruined barns, train cars, and old homesteads.

Not all of the residue from the past was as picturesque. Among the most pernicious leftovers were from salt brine spills that contaminated farmland for generations with the byproducts of drilling. Greenhouse gases were invisible, of course, but left a lasting impact in North Dakota and beyond. And then there was all that municipal debt, lingering long after the shine dimmed on boom times.

There with Krueger in the recreation center, I wondered what would happen. Williston would, once again, be paying off its debts for decades. But a bust was inevitable. Oil prices never stayed stable. And what about the bigger bust that was sure to come when we—I hoped— stopped digging up so many fossil fuels to keep ourselves warm?

The ARC wasn't even open yet, but walking its nearly finished hallways, I had a premonition of future emptiness. I could imagine tumbleweeds blowing down the corridors, past a toppled over Segway and empty swimming pools. Birds nested in the rafters of the natatorium, the waterslide all dried up. Coyotes peered around open doorways in search of rabbits. In my apocalyptic vision, I could even imagine mule deer bedding down on the artificial turf, nature reclaiming the prairie.

I shook off the sense of doom, and thanked Krueger for the tour. The ARC would be open the next time I visited Williston, and I promised to return for a whirl down the waterslide.

After the tour, I exited into bright sunshine and bitter cold. I wanted photos of the waterslide from the exterior, so I walked around the building. It was -17°F, a temperature that seemed incongruous with the bright sunlight. My ancient long underwear weren't much help. Worn so rarely, they dated to a time when higher-waisted jeans were in style. That meant the waistband bunched six inches above the top of my lower-rise jeans, which kept sliding down the slippery, synthetic fabric of the underwear.

I paused to hike up my jeans by my belt loops. And that's when I lost my footing on the ice. I slammed to the earth, tailbone first. I was uninjured but dazed. I looked around. No one saw me slip. No one was around to ask if I was okay or to offer any sympathy—it was too cold for people to be outside.

I hobbled back to the car, carefully. I started the engine, grateful for the seat warmer on my bruised bum. I caught a glimpse of myself in the mirror. After only five minutes outside, my cheeks matched my pink, fleece-lined hat. As soon as my fingers warmed up enough to use my phone, I googled "how long frostbite?" I got no definitive answers, and it was nothing compared to what oil field workers faced daily. But everything I read suggested it was wise to return to my warm car when I did.

———

I spent the night in a motel in Stanley, another oil patch town seventy miles from Williston and just sixty miles from Anna's homestead. The temperature dropped to -27°F. It was impossible to do anything outside except dash—carefully on ice—to and from the car.

Stanley was home to The Leader, a bar I had visited one afternoon a few months earlier. It was on the town's main drag, in a former newspaper office. It caught my eye while driving through the previous summer, and I had pulled over, thinking I could eat dinner there. But the bar didn't serve food yet—the owners told me they were trying to obtain a liquor license that allowed them to serve meals. The whole point of the piece I wanted to write was about how it was easier to get a permit to drill for oil than it was to open a small business in North Dakota. It was supposed to be one of those cheeky stories that points out incongruities in government policy. How small businesses get squeezed, but big ones have plenty of latitude.

The motel was little more than a hastily built box thrown together to house oil field workers, destined to be torn down or abandoned when guests no longer needed to stay there. It seemed as poorly insulated as some of the early homestead shacks. Wind whistled through the cracks around the ill-fitting windows. It was too drafty to work at the desk near the window, so I wrote up my story about McCarthy from the bed, my legs under the covers.

I was cold in my cheap hotel room and hungry. There was nowhere to eat at that hour in Stanley besides a Subway sandwich shop, allegedly the busiest per capita franchise in the company's chain. Hungry as I was, I couldn't bring myself to go out in the cold for a six-inch sub with wan lettuce and out-of-season tomatoes. The motel had a crockpot of soup downstairs in the lobby, but I was skeptical of the staff's food-handling skills or the hygiene of other guests who touched the ladle. Better to go to bed hungry, I decided, than risk food poisoning.

Early March in North Dakota was also when Anna lost her grip on her sanity. The winter of 1906–1907 was especially brutal in North Dakota. One of the women homesteaders I read about, a single school-teacher from Iowa named Lucy Goldthrope, told an interviewer that the winter of 1906–1907 was "the worst known up to that time in the Dakotas. From the middle of November it seems as if there was a blizzard about every other day until spring."[7]

Regardless of what she did to winterize her shack, "the cold crept in through the thin walls," Goldthrope said. The exterior was covered on the outside with tar paper. The interior was covered with blue building paper—ceiling and floors as well. She'd added gunny sacks over the paper and homemade wool rugs atop those. "The howling wind and driving snow, the mournful wail of coyotes searching the tormented land for food did nothing to make the winter any more pleasant," she said.

The state saw day after day of subzero temperatures and blinding snowstorms, said Scott Randolph, a business historian at the University of Redlands in Riverside, California.[8] I tracked Randolph down to learn more about the winter of 1906–1907, a time he studied because of its significance for railroad expansion in the West, one of his areas of expertise.

"It was bad for North Dakota people; it would have been disastrous anywhere else in the country," Randolph told me.

Because the harvest was late that year across the country, the railroad cars that transported grain to the East hadn't yet made it back filled up with coal. And when they tried to return, the snow was so bad that the tracks kept getting buried. It took days just to go twenty miles. Railroads struggled to supply the coal they needed just to fuel their own locomotives.

The newspaper accounts were exaggerated, Randolph said, but the coal shortage was not. No one died, but people were definitely hungry and cold. Many were trapped in tiny towns along the Soo Line railroad, the line that traveled to the northwest from Minot toward Anna's homestead. The towns were new, and because they were such young places, few had the sort of established institutions that offered a community safety net when times were hard.

People were angry in the aftermath, and they blamed the railroads. The Interstate Commerce Commission held hearings in Minneapolis and Chicago. Randolph told me there were real political consequences in North Dakota and elsewhere because of the bad weather, including a surge in progressive legislation and politics in subsequent years. That cold winter wasn't the only factor, but it was in part what led to cooperative ownership of grain elevators and a state-owned bank in North Dakota.

Anna and Andrew were relatively well off, but their place was small. Before her marriage, Anna had always lived with many other people. She grew up in a house full of siblings, and then as a young woman, lived with her sister and brother-in-law and her nieces and nephews in Fargo.

Her marriage to Andrew was her first winter in such a remote place. She would have been alone in a cold, small house with just her husband and a baby. From her medical records, I knew she was also bleeding heavily after her son's birth. The winter of 1907 might not have caused her undoing, but it surely didn't help.

The cold sapped me of all energy, including the confidence necessary to go to a bar alone as a woman in the oil patch on a Saturday night. I piled more blankets on the bed and decided to abandon the story about the bar.

Early the next morning, I checked out of the drafty motel and drove back toward Minot. I headed toward the mall, the only place I could think of where I could take a walk safely, out of the cold and around other people. I reconsidered my skepticism about the ARC in Williston. Maybe they needed a big recreation center so that people could walk around somewhere safe and warm in the company of other humans even when it was so cold.

As I drove from Stanley toward Minot, I once again felt that sense of foreboding that surfaced after my tour of the recreation center in Williston. It was so cold, yet so bright I had to don sunglasses to drive. Even the air looked barren and brittle. Drifts of dry snow blew like white dust across the road, sparkling in the blinding white light of winter. Possibilities seemed to be drying up all around me.

We could be rich, that insistent whisper down the generations, seemed also to blow away. The wind snatched up the words and scattered them across the plains.

11

THE MYSTERY BABY-MAKING BOULDER OF LIBERTY, ILLINOIS

September 2014

North Dakota crude: *$87.66 per barrel*

I COULDN'T BEAR ANOTHER LONELY JOURNEY TO NORTH DAKOTA BY myself, so I invited my husband to come along with me on my next road trip. We mapped a 2,000-mile route that would take us along roughly the same path as the Lewis and Clark expedition, across Nebraska and north through the Badlands of both North and South Dakota.

Chris can't drive, but he's skilled at finding restaurants and roadside attractions from the passenger seat. There was another consideration: it would be impossible to have a baby if Chris and I spent too much time apart.

As we approached the Mississippi River near Hannibal, Missouri, I asked Chris to poke around on his phone. Find us something quick and fun to do there, I told him. A writer ought to pay homage to Mark Twain, even if it is just to visit the town's Tom and Huck statue. I also wanted ice cream.

His online search unearthed something even more intriguing than Hannibal.

"There's the Mystery Baby-Making Boulder about a half an hour away," he told me.

"Go on," I said, laughing a little nervously.

Chris read out loud the description found on Roadside America, written thirteen years earlier by a man who said he lived in Liberty, Illinois: "It is a glaciated granite boulder, found in a local farmer's field, and dragged here as a conversation piece about 50 years ago. I have heard many stories about it—the most interesting is that long ago it was used by the Indians of the area in fertility rituals. Any new bride climbing up and sitting on the stone would become pregnant within a year! I can testify that it has worked for my wife and two other local ladies!"[1]

"Should we go?" I asked.

"That's up to you," Chris said. "You're driving."

The mystery boulder beckoned for a few exits. I was bored by the sameness of the day's drive, which had as its aim to cover as many miles of the Midwest as possible by sundown. We crossed Ohio, Indiana, and much of Illinois, where the only stop Chris could find with any charm was Abe's South Side diner in Springfield, an Abraham Lincoln–themed restaurant with a gravy-laden menu. Its specialty was the Horseshoe open-faced sandwich, a Springfield invention. The Horseshoe started with two slices of bread as a base, topped with a layer of meat, followed by hash browns, two eggs, and a cascade of sausage gravy. Chris ordered one, of course, if only to honor his own Midwestern roots.

We planned to spend the night in St. Joseph, Missouri, and the drive to Liberty would add at least another forty minutes to a nine-hour day of driving. As we drove, I considered what it meant to stop in Liberty—and not just the additional time on the road. Our journey from Washington, DC, to North Dakota was supposed to give me time to reclaim the story I set out to tell a year before, before the logistics of infertility overtook me. This journey was my chance to get back to Anna's story, back to the story of whether an oil company would drill on the land, back to North Dakota for more research. It was a chance, for a few weeks, to put aside all that yearning for a baby.

But a stop in Liberty meant acknowledging I was so desperate to get pregnant I would drive an hour round-trip to sit, in public, atop a phallic-shaped rock at a roadside attraction in rural Illinois.

I activated the turn signal and took the exit.

Our soundtrack on our road trip to North Dakota was the unabridged audio version of *Undaunted Courage*, Stephen Ambrose's biography of Meriwether Lewis and William Clark and their journey west from St. Louis and back again in 1804 and 1805.[2]

Chris had read *Undaunted Courage* a few years previously, but I had not. It never seemed necessary. The history of the expedition was in the air I breathed as a child in Oregon. I had absorbed it without trying, or so I thought. But until listening to the book on our drive, I'd never really considered how much of the journey took place in North Dakota. (Elwyn B. Robinson, in his history of the state, notes that although the Corps of Discovery's time in North Dakota was incidental to the mission as defined by President Thomas Jefferson, it "looms large in the history of the state. They spent more time in and made a fuller record of North Dakota than any other region through which they passed."[3])

Until then, I also hadn't thought much about the pivotal role of Sacagawea, the teen mom in a coerced marriage who accompanied the expedition from her home in North Dakota to Oregon and back. (In North Dakota, her name is spelled Sakakawea on maps and official documents.) I wondered what Anna knew of the Shoshone guide and translator, who in the early 1900s was the subject of renewed historical interest. Sacagawea had been taken up as a symbol of the suffrage movement, at the time of the expedition's centennial. Anna must have heard about efforts on the part of women's clubs in North Dakota to raise money for a statue. If the warden's words were true about allowing patients to listen to records on the Victrola and to read magazines and newspapers, Anna and others at the asylum would have seen the front page story in the *Jamestown Weekly Alert* about the statue's unveiling in October 1910.[4] The event drew five thousand people to Bismarck, a number that even now is a very large gathering in North Dakota for anything other than the state fair or a college football game.

In more recent years, Chris and a few of his river- and history-loving friends had retraced several portions of the historic expedition along the Missouri River through Montana. Some of them also re-created in canoes the final trip Lewis took down the Mississippi River, right before the explorer's mysterious death on the Natchez Trace in 1809. A river trip allowed them to experience travel at the same pace of the Corps of Discovery, albeit with modern conveniences.

In Jefferson's time, water travel was the fastest available method of transportation. "The Americans of 1801 had more gadgets, better weapons, a superior knowledge of geography, and other advantages over the ancients, but they could not move goods or themselves or information by land or water any faster than had the Greeks or Romans," the voice-over actor Barrett Whitener intoned as we sped west.[5]

But sixty years after the Corps of Discovery ventured west to Oregon and back again, generals fought the Civil War at the speed of telegraphs and trains. Chris and I were, of course, blowing by exits at seventy miles per hour or more as we left Washington, DC, and drove into western Maryland and then on to Pennsylvania and Ohio along the path of the C&O Canal. It took us three days to cross nearly halfway across the continent in a car, and we could have done it in mere hours on a plane.

We were hardly on a grand adventure into unknown territory— although neither one of us had ever been to Nebraska. A road trip, though, was exactly what our marriage needed: time with each other in the car to sort through what we wanted next in our lives. All travelers seek transformation of some sort, whether it's riches or a new understanding of the world or time to separate ourselves from the daily grind of our ordinary routines. Otherwise, we would just stay home. Otherwise, we wouldn't take the exit to Liberty.

It was Labor Day weekend, a sunny, bright end-of-summer afternoon perfect for picnics and barbecues. We exited the highway and drove toward Liberty through rolling hills and corn fields, barely encountering another car. The streets of Liberty were just as deserted, until we got to the town square. Cars filled the gravel lot surrounding the scruffy green. Just about everyone in town and from the surrounding countryside seemed to be picnicking there.

I drove slowly, the car tires crunching the gravel as I sized up the Mystery Baby-Making Boulder. Definitely phallic, the boulder looked like the deformed, squat tip of a penis. It was about hip high, a dirty tan. I could perch on it, like a frog on a toadstool.

To get to it, though, I needed to walk through the outfield of a multigenerational whiffle ball game. I parked, wondering whether I could actually summon up the courage to walk to the rock in front of all of those people. To do it meant they would know what I wanted.

I exited the car and walked toward the outfield. I asked one of the women if the rumors about the rock were true. No idea, she said, she was just visiting. And she had zero interest in sitting on it herself. No more kids for her, she told me.

Mortified, I imagined how she pictured me: a woman so desperate she drove to Liberty on Labor Day weekend in a car with East Coast plates. A woman who was striving for something she should have thought about a decade earlier, when it wasn't pushing the edge of too late.

Avoiding eye contact with anyone else in the field, I held my head up high and walked straight toward the rock. I perched atop the boulder in a blue-and-white striped maxi dress. I posed pinup girl style, grinning. Chris snapped my picture, amused by my discomfort. I thought of his jaunty walk every time he exited the fertility clinic to do his part. Chris didn't care what anyone thought; he just did whatever needed to be done.

Perched on that midwestern boulder, I, too, stopped caring what anyone thought of me for sitting on that rock. In the photo, there's no fear in my face or my posture, only happiness. There's pure joy in the picnic perfection of the day, the sweetness of my husband's company, and the way we embraced the goofy side trip. We had ventured off the beaten path, and if the boulder ended up working, the story would, in turn, embarrass our child for decades.

The Mystery Baby-Making Boulder of Liberty, Illinois.

It was liberating to admit it: I was a forty-one-year-old woman who really wanted a child. I wanted it so badly that I would drive well out of my way to sit on a penis-shaped rock in the Midwest just in case it actually did have mystery baby-making powers.

Giddy, I drove us to Hannibal in search of huckleberry ice cream. As we left Liberty, children clambered over the rock. One little girl even posed like me.

———

All trip long, I collected highway names for our possible child. It started as a family joke in 2008, when my sister was pregnant with twins. At the time, our mother was in a nursing home in Oregon, recovering from one of her heart attacks. We drove daily along Interstate 5 in my dad's minivan, through the flattest part of the Willamette Valley from Salem to Portland and back. My father and I suggested terrible names for the twins, culled from exit signs and overpasses: Donald and Aurora, Brooks and Gervais. Waconda and Chemawa. And my favorites: Quinaby and Concomly, both roads.

Six years later, driving through Illinois, I told Chris about the family game. We joked about Midwest-inspired possibilities for our own future child. Hannibal was out, obviously. But Liberty had its charms, especially if the fertility rock worked. And Ohio, Indiana, and Illinois brought us Dickensian exit sign names suitable for characters I might write in a novel or a screenplay one day. I asked Chris to enter the names in a file in my phone while I drove: Fithian and Rankin, Cisco and Weldon, Oreana and Decatur, Macon and Brookfield.

The highway signage reflected the arrival of explorers and settlers. They arrived in these places, naming them after themselves or loved ones, or leaders they admired or patrons they wanted to flatter. They named places after people they pined for, or the places they came from that they might not see again. Sometimes, they anglicized Indigenous names. Often, they erased entirely the presence of the people who came before them by assigning a place a new name. Always, their arrival changed these places. Their presence made a mark, and not just on the map. Their arrival immediately altered it. Consider the Dakota Access Pipeline, a name a Texas energy company had just given to a proposed pipeline that was about to explode in the public consciousness in Standing Rock, North Dakota. "Dakota is our name. It means allies, friends," Faith Spotted Eagle, a Yankton Dakota elder told the writer Jacqueline Keeler for her book, *Standoff.* "How can they use it for their pipeline? They are not being allies to us or our Mother Earth."[6]

Speeding by on the interstate, there was something unsettling about how little of a mark Chris and I made. The haze of our car exhaust on the horizon, the flush of a toilet at a rest stop. A credit card receipt for gas.

I knew as a journalist that even the most mundane-seeming places had a story. And it was what kept drawing me back to North Dakota, a place so overlooked by outsiders that the tourism office had a "Best For Last Club" that gave out T-shirts to people who made the state their final destination in a quest to visit all fifty states.[7]

On our first leg of the journey, though, it felt as though Chris and I could be anywhere. The anonymity was unsettling. Earlier that morning, we had stopped for coffee at a Starbucks on the edge of Indiana, in a shopping center right off the highway. The drive-through customers got priority. As customers inside the coffee shop, our order took forever.

I experienced a moment of dislocation. The layout was identical to every Starbucks I'd ever been to in suburban America—everywhere but seemingly nowhere. The notice board included the same sort of community announcements tacked up at all coffee shops: business cards from the Mary Kay rep and the local insurance agencies, notices for kitten adoption fairs, and bootcamp fitness classes. I felt as though I had been there before, even though I knew it was impossible. I didn't even drink coffee in 1994, the last time I could remember driving through Indiana.

"Wait," I thought. "Where am I?"

That question lingered all day, temporarily shoved aside when we sidetracked to Liberty and Hannibal. But as we made our way into St. Joseph, Missouri, it reemerged. Our motel, paid for with credit card miles, was a chain on the highway, next to a shopping center anchored by a Target. We bought some camping supplies at the store, in what could be any suburb in America. We bickered about where to get dinner. Nothing was open on the holiday weekend but the chain restaurants near our motel, and I began to fear we would have to eat at Applebee's. We were just another hungry couple in an American suburb, in search of a decent meal. We could be anywhere off any U.S. highway.

Sleepy as it seemed that weekend, plenty had once happened in St. Joseph.[8] It was a gateway for westward expansion, a Missouri River town that began as a fur trading post, and a place that thrived as prospectors hustled through on their way to California and along the Oregon Trail. In 1859, St. Joseph was the second-largest city in the state, smaller only than St. Louis. It served as the eastern terminus of the short-lived Pony Express, and as railroads crisscrossed the U.S., it emerged as a wealthy depot on the Missouri River. St. Joseph in 1882 was where a fellow gang member seeking a reward shot to death Jesse James in a home where the infamous outlaw was hiding out under an assumed name.

In search of a meal, Chris and I drove toward downtown along Lovers Lane, made famous in the nostalgic poem by the newspaperman, Eugene Field, a St. Joseph journalist who was celebrated for sentimental children's poetry, including "Little Boy Blue" and "Wynken, Blynken and Nod." We wandered around the riverfront and downtown, its fading beauty more glorious in the waning light. I tried and failed with my camera to adequately capture the scale of an Oregon Trail–themed, multistory mural painted on the back of three downtown buildings.

I couldn't imagine anyone ever making the effort some future day to paint such a mural on a decaying suburban Target. Somewhere along the way we'd embraced the efficiency of consistency over the messiness of originality. Maybe, I thought, this was one of the reasons why many of us want children. Maybe it's the most creative thing we can do in a world where the chain restaurants with predictable Blackened Chicken Caesar Salads are the only thing open on the Sunday night before Labor Day. A child is a lasting mark upon the world. A messy and inefficient mark, but a mark that no one else could make in quite the same way. In a world where all the suburban shopping malls and all the interstate Starbucks were the same, a child was a unique creation.

That night, I went to bed dispirited, uncertain of my own mark on the world. I'd quit my job to pursue Anna's story, but the logistics of fertility ensnared me in their inescapable truth: I had to act quickly or not at all. I'd already undergone a surgery to fix a blocked fallopian tube. We'd already tried one unsuccessful round of intrauterine insemination, known as an IUI. If the fertility rock didn't work, we faced more aggressive intervention when we returned home.

I also missed the daily thrum of the news business. I felt disconnected from the news cycles that once drove most days and weeks of my adult life. On the other side of Missouri, people were protesting the death of Michael Brown, the eighteen-year-old killed by police in Ferguson, a suburb of St. Louis.[9] Ferguson was named in the 1850s for a farmer who deeded to the railroad a strip of land running through his farm, in exchange for a depot. The farmer was long forgotten, but the weight of events in the town named for him birthed the Black Lives Matter movement. It also transformed Michael Brown's name into a symbol of the fight for racial justice in America. Names carried weight.

That night, I read up on Field, our sentimental poet. He was the son of the lawyer who represented Dred Scott, the enslaved man whose suit in Missouri federal court was among the events that led to the Civil War.

I couldn't square the juxtaposition of history with what was going on in modern-day Missouri that summer. It seemed we'd abandoned much of the beauty and singularity of the past, even as we carried forward the worst of it. What would it mean to bring a child into a world where we had no eye for justice?

The next morning, I woke up early to run through the shopping center surrounding our hotel. We had another big day of driving ahead of us, and I wanted some movement before I got back behind the wheel.

Unexpectedly, there were sidewalks circling the development near the hotel. It seemed unlikely anyone would walk there but I followed one of the sidewalks toward a retention pond and a wooden bridge. It was perhaps the most picturesque use of stormwater runoff I'd ever seen in a suburban shopping center. Mid-run, I stopped on the bridge, struck by what I saw when no one else was around: a blue heron, looking for breakfast. The bird waded into the water, in search of small fish, frogs, or other reptiles. A few ducks made lazy circles around the pond. The silence was interrupted only by an unheeded security alarm blaring from the Chick-fil-A at the edge of the shopping center.

I ran around the pond three times, trying unsuccessfully to photograph the heron with my phone. I finally gave up, laughing out loud at myself for all the effort. I was running in circles, chasing the ephemeral, documenting something that was best experienced. Even though

I wrote off the suburban development for its soul-crushing anonymity, there it was: beauty. At a retention pond in a shopping center off the highway in the middle of the country.

On the run back to the hotel, I noticed a Starbucks across the street from the shopping center. There was a drive-through. It was right near the highway on-ramp. The coffee would fuel us through the sandhills of Nebraska and on to the Badlands of South Dakota.

This time, we chose the drive-through. It's reassuring to have predictable coffee. Lewis and Clark started their journey with fifty pounds of coffee beans, after all.[10]

12

BEATRICE

September 2014

North Dakota crude: *$87.66 per barrel*

JUST OUTSIDE THE SMALL TOWN OF BEATRICE IN SOUTHEAST NEBRASKA IS Homestead National Historic Park, a place well off the beaten path. The town's name is pronounced with the emphasis on the middle syllable: Bee-AT-tris. I learned this before our arrival from my friend, Noelle, who once worked a summer internship as a park ranger there. Nebraska lore held that the town was named in 1857 for one of its early white settlers, Julia Beatrice Kinney, a teenager who enjoyed making people puzzle out the unconventional pronunciation of her middle name.[1]

Beatrice was another cute entry for our baby name list. Baby Bee. I liked it, although perhaps with the more typical emphasis on the last syllable: Bey-uh-treece.

The streets of Beatrice were quiet as we drove to the visitor center at the monument. The building, with a swooping roofline designed to evoke the image of a plow breaking through sod, sits amid one hundred acres of tall grass prairie. The Park Service restored the surrounding grasslands so that visitors could see what the prairies would have looked like in the mid-1800s.

The terrain was not unlike Anna's homestead more than eight hundred miles to the north, although there were fewer of the pothole lakes that marked North Dakota and many more trees. The prairies were dotted with sunflowers and tall purple thistles, the grasses faded to a dry, late summer green. Scrubby lowland bur oak woodlands marked the creek bottoms, serving as valuable windbreak and shade on the open prairies as well as a source of firewood for settlers seeking homesteads.

The shade trees and the water source were exactly what drew Daniel Freeman, one of the nation's first official homesteaders, and they were the reason the monument was in Nebraska. Freeman, a Civil War veteran born in Ohio, filed his homestead claim in the early hours of January 1, 1863, the day the law took effect. Two years later, he persuaded a young woman from Iowa, Agnes Suiter, to join him in Beatrice. (Agnes was once engaged to Daniel Freeman's brother James, who died in the Civil War.) Eventually, the Freemans had eight children. They lived prosperous lives. Some of their children married and built homes of their own on the original homestead.[2]

I wandered through the exhibits in the visitor center, reading the interpretive signage.

Many Americans in the 19th century believed in Manifest Destiny, the idea that the United States had a God-given duty to expand westward and tame the continent. The Homestead Act embodied this belief.

In an adjacent exhibit, the curators posed a question that might have challenged the understanding some visitors had of their ancestors.

Whose land was it? What does it mean to own the land?

Settlers and American Indians each answered that question differently. Many homesteaders believed that native people were nomads, and that only those who owned land would use it efficiently. Tribes such as Nebraska's Otoe-Missouria, however, were not nomadic. For most American Indians, land belonged to the community, not to individuals. They didn't own the land the way homesteaders conceived of ownership.

I couldn't decide whether the messaging was muddled, or if the Park Service was trying—not always skillfully—to balance multiple truths about the complexity of our homesteading past. (And in fact, in subsequent years, the interpretive language has grown more nuanced, especially around the displacement of Indigenous people.)

I did appreciate how the curators chose to highlight the stories of individual homesteaders like Freeman, a decision that made vivid the lives of the people who built homes, families, and futures on 160 acres of nearly free land.

The historic park also highlighted the life of Oscar Micheaux, among the best known of the Black homesteaders who made claims in the Great Plains. Micheaux was born in Illinois in 1884 to parents who began their lives in slavery. He traveled the country on the railroads as a Pullman porter. Like so many homesteaders whose lives are featured in biographies published by the National Park Service, "the prairie called to him." Micheaux homesteaded when land became available on the Rosebud Indian Reservation in South Dakota in 1904 and stayed on until a three-year drought ended his time as a farmer.[3]

Micheaux wrote two books about his time in South Dakota including *The Homesteader,* which caught the attention of an early film production company. In 1919, Micheaux adapted *The Homesteader* into a movie, becoming the first known African American filmmaker and director. The film is lost to time, like many of the era, but other

Micheaux movies survive. He went on to make forty films over thirty years, including *Within Our Gates* in 1919, considered a Black rebuttal to the racism of D. W. Griffith's *Birth of Nation* in 1915.

Many Black homesteaders formed communities of fellow African American settlers, most notably in Oklahoma but also in Colorado, Kansas, Nebraska, New Mexico, and Wyoming. But the communities were exceptions rather than the rule. It was difficult for most Black settlers to make a life in the West.[4]

My home state of Oregon, for example, had explicit exclusionary laws on the books before statehood and was a particularly hostile place for Black Americans who later wanted to take advantage of homestead laws. In 1844, Oregon's territorial legislature banned slavery, but also prohibited Black people from residing in the territory for more than three years. Another law in 1849 explicitly barred Black people who weren't already in Oregon from visiting or residing in the state. At statehood in 1859, lawmakers who wrote the Oregon constitution doubled down on exclusionary practices by prohibiting Black people from owning property or making contracts. Although the laws were rarely enforced and later repealed, they nonetheless did exactly what their authors intended after the Civil War: They discouraged Black Americans from taking advantage of the Homestead Act in Oregon.[5]

Many white visitors to Beatrice probably found it difficult to wrap their heads around the idea that those who came before them had a role in displacing the original inhabitants of the land, or in discouraging Black settlers. We aren't responsible for what our ancestors did— consider the man I read about in Fergus Falls who threw a woman's body in a pigpen in Minnesota in 1887! But the systems those ancestors created persist in their inequality, and those of us alive can do something about creating a more just society to live in now and in the future.

Yet even in Beatrice, a place devoted to such questions, I wasn't sure many visitors gave the systems of the past much thought. Like the genealogy researchers in Fergus Falls, I suspected most people who visited were in search of stories that validated their status in the present, not those tales exposing the faults of their forebears or the systems they created and that continue to benefit their descendants.

Not that there were many people to ask. Although the historic park in 2012 drew more than 100,000 people for special events marking the 150th anniversary of the signing of the Homestead Act, that was a big year for visitors. Most years, only about 80,000 people visit. It's a quiet corner of a quiet state, at least forty miles from an interstate highway, and far from regional population centers like Kansas City or Omaha.

As I continued to wander through the exhibits, one of the interpretive signs made me pause. I laughed out loud:

> *Few homesteaders grew rich. But many felt rich with satisfaction. They felt a glow of achievement in having land to pass on to their children.*

This seemed an overly optimistic assessment of human nature as I knew it, shaped by American capitalism. "Rich with satisfaction" was not a whisper that would have compelled most Americans of the homesteading period westward across the prairies.

Yet it is also true that the Homestead Act is one of the world's most successful experiments in wealth redistribution.[6] It lifted countless people, mostly white families, out of poverty by making them land owners. An estimated forty-six to ninety-three million Americans, myself included, are descended from the people who successfully claimed 1.5 million original homesteads.[7] I wondered how many of those people were getting rich off mineral rights in oil patch states like North Dakota and Oklahoma—it seemed like a project for an ambitious doctoral student with superstar GIS mapping skills and access to data at Ancestry.com.

Just like my mother, many people must have been surprised to get manila envelopes in the mail offering them leases. You could, I thought, hold two truths at once. You could admire the hard work and sacrifice of your ancestors, even as you understood how the original homesteaders began building generational wealth that excluded so many Americans from such opportunities. The next step, of course, is to remedy the inequity in the present and the future.

We walked outside to one of the cabins, built in 1867 about fourteen miles northeast of the park and trucked in in 1950. Chris and I posed for a photo in the entrance of the cabin, which was about the dimensions of our bedroom at home. The cabin was 14x16, the same size as one of the two cabins my great-grandfather Andrew claimed on his land in 1912. But unlike Andrew, who was the sole inhabitant after Anna's exile to the asylum, ten children and two adults lived in the 224-square-foot Nebraska version. It was difficult to imagine so many people living there.

Homestead National Historic Park.

We left Beatrice, reluctantly. It was a place I could have lingered for hours while wandering the rehabilitated prairie, conjuring up fantasies about Anna's brief time as a homesteader. But we needed to be on the road to Badlands National Park in South Dakota in time for Chris to set up our tent before dark. This division of labor was informal but mutually agreed upon: I did the driving, so Chris set up the tent and made dinner.

———

The next day, I spotted a herd of bighorn sheep on a narrow outcropping beneath a ridge. It was about halfway into our hike. The angle of the light was beginning to hint at the shift toward wintertime, but it was still eighty-five degrees mid-afternoon on a cloudless day, and the herd, all bachelors, eked out the sole shade in a treeless wilderness.

Some of the rams sat, some stood watch. Their coloring was the same tawny color as the geology, their coats camouflage. My eyes at that point in the hike were primed for spotting animals, although I was mostly on alert for wildlife closer to the ground. Signs throughout Badlands National Park warned of rattlesnakes, and other hikers told us earlier that they saw snakes on a separate short trail.

I led the way. Chris has the reflexes of the bachelor rams, and he moves forward into the world with a purposeful gait I can spot a block away. As nimble as Chris is on his feet, his vision is limited. I'm the one in the marriage responsible for spotting snakes.

From afar, the sharp relief of the formations against the blue sky gave the Badlands the appearance of something forever and unchanging. Up close, though, they crumbled like streusel topping. It was an unstable scree, best suited for sure-footed ungulates. We were intruders here, and the instability of a landscape shaped by wind, water, and extreme temperature offered warnings: nothing is permanent.

We paused so Chris could see the bighorns through his binoculars. When pressed, Chris will tell people he doesn't see very well. If he can get away with it, he won't say anything at all. People who know him only casually don't know he is legally blind. His mother once told me that when Chris was a child, she caught him up on the counter in the bathroom, peering at his eyes in the mirror. He asked her why his eyes were so different from everyone else's. It was a rare moment of self-pity about his vision, one I have never heard him express.

As one of six kids, little held Chris back. He did almost everything his older siblings did. He rode bikes and played youth soccer in his neighborhood, played trombone in his high school marching band, and, unlike his brothers and sister, he was allowed to sit close to the television for Saturday morning cartoons. He told me once that when it was his turn at Jarts, the lawn dart game popular in the early 80s and since banned by the Consumer Product Safety Commission, his friends pointed their feet at the target, and then swiftly removed their toes as Chris tossed.

Some times of day are challenging, like dusk, when the angle and low intensity of the light make it harder for Chris to see. That's when he knocks over wine glasses or walks into low-hanging tree limbs and crooked signposts. He also has trouble reading signs at restaurants or in train stations and airports. Although he has a superb sense of direction, he can be an irritating navigator because he doesn't see the road signs or exits ahead.

"It should be right here," he often says.

"Where?!" I will ask, frustrated.

The hike in the Badlands marked the one-year anniversary of leaving my job to pursue Anna's story. Here I was, in a raw, wild, and wide-open space, the sun beating down on me, and the weathered rock formations imprinting their shapes into my soul. The dry heat and mild breeze evaporated the sweat from my skin, leaving a salty, sunscreen grit behind. I scaled to the top of a boulder to get a wider view of the terrain. I had no idea what came next. But that day, it didn't matter. I was exactly where I wanted to be.

A few hours later I limped toward the car, hobbled by what turned out to be among the worst blisters of my life. I removed my sweaty hiking shoes. Skin peeled away with my bloodied socks, leaving me not only gasping with the pain, but oozing blood and pus from the blisters. I slid my feet into my flip flops, the air painful on the open wounds of my heels.

We leaned against the trunk, downing the cold beer Chris stashed in a cooler earlier that morning. This was another of his skills—anticipating the perfection of a cold beer after a long, hot hike. The beer seemed to cool my body from the inside, down my belly and legs, all the way to my feet. The mild buzz eased the pain of my blisters. We felt the relief of having completed a ten-mile hike in the heat and the pleasure of rewarding ourselves for our work. It felt like we were searing the otherworldly landscape not just into our brains and memories, but into our bodies.

Chris and I sometimes talked about how we wish we met when we were younger. We didn't have this conversation often, because it wasn't helpful to dwell on something we couldn't change. We both knew that children were not a guarantee for people who met at nearly thirty-eight and married at thirty-nine, and we were realistic about our chances. It wasn't something we could explain without sounding cloying or sentimental, but in our minds, meeting later in life meant that we were more devoted to how we chose to spend the rest of our lives together. We had less time left to get it right.

That night in the South Dakota Badlands, we watched the sun go down over the rock formations. Fires weren't allowed in such an arid place, so we ate freeze-dried camp food boiled in a bag on our gas stove. That night, we drank cool white wine and talked about what sort of parents we wanted to be. We definitely wanted to be the sort who took a future child camping in the Badlands. People all over the world were having such conversations about their futures. Our yearning was universal, and this was comforting.

Chris asked me to explain something I did as a journalist. How, he asked, do you experience things and witness events and then turn it all into something readable and understandable. "How do you do that?" he asked. "How do you know what to write?"

I thought about it for a moment. The wine made it seem so simple, even though it is not, not at all. I borrowed a line from my friend, Jill, who worked in communications. "You have to know what you want to say," I told Chris.

It was a truth that held up in the sober light of day, as we packed up our gear and headed toward North Dakota.

I finally understood why I felt so ashamed in Liberty at that fertility rock. I knew what it was I wanted to say. We were admitting, by stopping in Liberty, that Chris and I wanted something that might be beyond our reach. Something we might not get to have. Our yearning was out in the open. It was nakedly apparent. If it didn't happen, other people would know about our disappointment. Other people might label it a failure. They would feel sorry for us when we didn't want them to. It would shape how they saw us, which would not be the same way we saw ourselves.

Neither one of us understood it quite yet, but we sensed it. We were about to cross a threshold of adulthood with no return. If we didn't have a child, the absence would be present with us the rest of our lives, something we yearned for and couldn't achieve, in lives of relative privilege with few limits.

The absence would be like grief. A sadness that retreats with time, but never really disappears. It would also be something we shared, even if we didn't share a child. Exposing our yearning to the world threatened the purity of our eventual, shared grief. I didn't want to expose this vulnerability to a world that didn't understand.

It may have been unsophisticated to draw inspiration from an interpretive sign at an obscure national historic park in Nebraska. But like the homesteaders who sought and failed to achieve material wealth, we, too, would need to seek out lives rich with satisfaction.

13

MEDORA

September 2014

North Dakota crude: *$87.66 per barrel*

SUMMER SEEMED TO RETREAT BEHIND US AS WE LEFT THE ARID HEAT OF South Dakota for the Badlands of North Dakota. As we approached the town of Medora, I pulled over at the Painted Canyon exit overlooking Theodore Roosevelt National Park. From the overlook, we could see the buttes, hoodoos, and pinnacles of the North Dakota Badlands. Softened by the valley of the shallow Little Missouri River and the surrounding grasslands, the formations appeared greener and rounder than the harsh, arid South Dakota Badlands. In the distance, a herd of bison roamed the park. A few miles away was Medora, the storied cowboy town of the West, now home to a pageant musical set each summer in an amphitheater with a view of the Badlands.

I had been to Medora the year before, to write about the effect of a 2013 government shutdown on national parks. Theodore Roosevelt National Park was closed, but I was able to drive through the unguarded gates. Emboldened by the lack of human activity, wildlife had the run of their own habitat. Within minutes of entering the closed park, I saw a coyote walking across a ridge. As I eased my car around bends in the road, I surprised families of pheasants. A herd of mule deer blocked my path, glaring at me like the intruder I was as they crossed the road in front of me. A flock of twenty wild turkeys waddled in a meadow that led to the Little Missouri River, their awkward neck wattles bobbing ahead of their bodies.

The Medora Musical was dark for the winter when I was in town. But because it was the shoulder season, I got a discounted rate at the fanciest lodging in town: the Rough Riders Hotel. Inside my room was a framed page from an old edition of the *Bad Lands Cow Boy*, the short-lived newspaper that documented Medora's heyday in the mid-1880s. I spent hours online reading old editions of the *Cow Boy*, inhaling the myths of the West in a primary source. The paper's founder, Arthur T. Packard, was a recent University of Michigan graduate and the son of an Indiana congressman and newspaper publisher. The first edition of his paper on February 7, 1884, spelled out his journalistic intentions in Dakota cattle country. "We do not come as the agent or tool of any man or set of men," he wrote. "There is a wide field for us to cover, and we intend to cover it. We do come, however, to make some almighty dollars."[1]

Ah, I thought, as I read Packard's mission statement: *We could be rich.*

Medora, it seemed, was one of the sources of the whispers blowing across the prairies, if not a motherlode. I knew then that I had to return in the summer for the musical, if only to see on stage, there in an amphitheater in the Badlands, how the myths of the West were manifesting in modern-day North Dakota.

There was plenty of human history in western Dakota in the centuries before a newspaper like the *Bad Lands Cow Boy* was there to document it. But it was the arrival of a rich, young French aristocrat known as the Marquis de Morès that established Medora on the map—at least for a time. The marquis arrived in 1883, the same year the Northern Pacific Railroad completed its transcontinental route. He named the town he established after his wife, Medora Von Hoffman, the daughter of a New York banking family. De Morès and his wealthy father-in-law, a railroad investor, snapped up six square miles of Little Missouri River bottom to build a town and a $250,000 slaughterhouse next to the tracks. Their plan was to send fresh beef to the east on refrigerated cars, without the expense of sending live cattle to stockyards and to slaughter in Chicago. De Morès hired 150 cowboys, and he purchased six thousand cattle and fifteen thousand sheep. All those sheep required barbed-wire fencing, an installation that portended the end of the open range, and angered the hunters and ranchers who had arrived in the Badlands before de Morès.[2]

Soon after de Morès arrived, so did another young aristocrat: Theodore Roosevelt. The future president of the United States showed up in Medora in the fall of 1883 to hunt bison. He was twenty-five years old, just like de Morès, and although Roosevelt was wealthy enough to invest in two ranches in his time in Dakota Territory, he was nowhere near as rich as a marquis with access to his father-in-law's line of credit. Yet within days, Roosevelt was seduced by cowboy life and the promise of ranching wealth. "All who visited the country were enthusiastic," wrote Elwyn B. Robinson in his history of North Dakota, "but none more than Theodore Roosevelt."[3] In his biography of the eventual president's rise, Edmund Morris was even more evocative of the vigorous, testosterone-fueled atmosphere of 1883 Medora: "A spirit of lusty optimism pervaded the place."[4]

In Medora, de Morès and his wife built a twenty-six-room wooden chateau atop a hill overlooking the abattoir. The marquis and his wife were young, good-looking, and rich, and they liked to hunt; Medora was reportedly an excellent shot from a sidesaddle. Her husband had a military background and was prone to feuds; before his arrival in Medora, he had killed two Frenchmen in duels. After he rode into Dakota Territory, de Morès faced multiple sensational trials and acquittals for his involvement in an ambush that led to the death of one of the hunters he angered with his barbed-wire fencing.[5]

Long before the spectacle of a live show in an amphitheater with a stunning view, de Morès knew the Badlands were an alluring travel destination for those eager to experience the authentic West—and those who might be persuaded to invest in his ventures. One of the front rooms on the first floor of the chateau was a working tack and equipment room. Its placement in the main house was deliberate, interpretive guides tell modern visitors. Dakota cowboys were a novelty; chateau visitors could see, hear, and even smell the working cowboys for themselves. "Daring, laughter, endurance—these were what I saw upon the countenance of the cow-boys," wrote Owen Wister in *The Virginian*, the 1902 novel often credited as the first of the Western genre. "For something about them, and the idea of them, smote my American heart, and I have never forgotten it, nor ever shall, as long as I live."[6]

Long before the publication of Wister's book, the pages of the *Bad Lands Cow Boy* newspaper crackled with admiration for life on the range and the promise of cattle riches. Even before *The Virginian* or the rise of western artists such as Frederic Remington,[7] it was as though Packard understood he was chronicling a brief moment of time in the West that would shape American culture for the next 150 years. "We have been asked why we named our paper the *Bad Lands Cow Boy*," Packard wrote in the second edition of his newspaper. "We adopted the name in the first place to attract attention, which it certainly does."[8] It's no coincidence that 1883 was also the start of Buffalo Bill Cody's Wild West Show, another form of popular entertainment that went on to shape the myths of the West for several decades and beyond. Cowboys were cast in Packard's paper as chivalrous knights of the Great Plains who had been unfairly drawn as outlaws. (Just like the masthead of the newspaper, Packard spelled "cowboy" throughout his pages as two words.) "The term 'cow boy' has been a reproach long enough," Packard wrote. "Every other paper in the land has joined hands to heap contumely on the head of the cow boy. We will stand singly and alone and uphold a name which we know represents a good cause."[9]

Roosevelt also took up the cause of the cowboy in the general interest magazine articles and books he wrote about Medora ranch life. And Wister's own book was influenced by his travels in Wyoming in the 1880s; his acquaintance with Roosevelt dated to their time together at Harvard. Years later, Wister wrote in a memoir of their friendship that he thought the future president was "the pioneer in taking the cowboy seriously," adding that he "loved what he said about that bold horseman of the Plains."[10]

"Roosevelt threw a charm over the Badlands in much the way other American writers have over other specific locales," wrote historian David McCullough, in his chapter on Medora in *Brave Companions: Portraits in History*. "And it was the Badlands cowboy, the rough rider as Roosevelt found him in and around Medora, that he fixed in the public imagination."[11]

Less than a week before Roosevelt's arrival in 1883 to hunt buffalo, white hunters and a band of Sioux killed off a herd of ten thousand bison near Standing Rock, with full approval of federal troops stationed in the region.[12] Vic Smith, the "most famous hunter in Montana," was among a "host of white hunters" who "took part in the killing of this last ten thousand," wrote William Hornady, then the superintendent of the National Zoo. "When we got through the hunt there was not a hoof left,"

Smith told Hornaday. Little remained of herds that once numbered more than thirty million.[13] The federal government knew that "plains bare of buffalo would soon be bare of Indians too," Morris wrote.[14]

This may not have been an especially alarming prospect to Roosevelt, who just a few years later during a lecture in New York, revealed his racist beliefs about the Indigenous people of the Northern Plains—and his conviction in the superiority of the cowboy way of life: "I don't go so far as to think that the only good Indians are the dead Indians, but I believe nine out of every ten are, and I shouldn't like to inquire too closely into the case of the tenth," he said in 1886. "The most vicious cowboy has more moral principle than the average Indian."[15]

Roosevelt's bison hunt, which he documented in minute detail in his own account, *The Lordly Buffalo*, was a grim search for one of the remaining stragglers from the decimated herds. Yet even as Roosevelt participated in the destruction of what drew him to the Dakotas, he fell in love with the Badlands, wrote Morris. "In recent years, he had spent much of his time in crowded, noisy rooms. Here he could gallop in any direction, for as long as he liked, and not see a single human being."[16] Roosevelt returned to Medora in 1884 after the death of his mother and his first wife, seeking solace in the landscape. He kept up his life in New York, but returned to Medora repeatedly for several years, until the winter of 1886–1887 dashed his dreams of ranching riches. The winter arrived unusually cold and early and lingered long—it was the year before the fatal storms documented in Laskin's *The Children's Blizzard*. Cattle starved by the thousands. The de Morès slaughterhouse closed. Fortunes went bust, including Roosevelt's cattle operation. Even the offices of the *Bad Lands Cow Boy* burned down, and Packard moved on, eventually becoming a well-known baseball writer.

The marquis returned to France with a diminished yet still considerable fortune, despondent over his poor showing as a businessman in the American West. Today, what remains of the de Morès vision of Medora is the brick chimney of the abattoir, and the restored chateau—Robinson called the mansion "a monument to an emotion."[17] Both are part of a state park and interpretive museum. The National Park Service describes the ruins as "a quiet reminder of the Marquis's unfulfilled dreams."[18] The entire place, including the musical in the amphitheater, might be more aptly described as a monument to a failed slaughterhouse owned by a megalomaniacal Frenchman whose dream of riches in the American West led so many to believe they, too, could prosper as cattle ranchers in the Badlands.

Roosevelt was a New Yorker through and through, but in Medora, he's considered a North Dakotan first. And it's true that North Dakota left a lasting impression on Roosevelt, long after the brutal winter dashed his cowboy dreams. It was during his time in North Dakota that Roosevelt continued to refine the conservation ethos that led to the system of American national parks and monuments.[19] It was also Roosevelt's time as a Dakota cowboy, most scholars assert, that toughened an unseasoned young man into the sort of an adult capable of one day leading the nation. "I have always said I never would have been President if it had not been for my experiences in North Dakota," Roosevelt wrote to the historian Albert Volwiler in 1918. It's a widely quoted phrase in North Dakota, a place far from major cultural and population centers, but eager for validation of its national influence.[20]

After the grim winter, Medora dried up into a near ghost town, not unlike many other boom towns of the West. Even now, only 134 people live in Medora year round. It got a small boost in the 1930s, when the Civilian Conservation Corps restored the de Morès chateau and built much of the infrastructure for the state park and an eventual national park.

For decades, though, Medora was just another example of North Dakota's boom and bust past. It was a dusty, downtrodden place lacking even sewers or a water system until a wealthy North Dakota businessman named Harold Schafer in his retirement began buying up buildings and restoring the town. It was Schafer who in 1965 started the Medora Musical in its modern iteration as well as a foundation that owns most of the lodging and businesses in town. Over time, he built it into one of North Dakota's most notable tourist attractions.

Schafer was the founder of the Gold Seal Company, which sold—but did not directly manufacture—cleaning products, including Glass Wax and Mr. Bubble. He began the company as a traveling salesman, and built his fortune with a keen, midcentury understanding of the power of advertising.

Schafer had long been enamored with the Badlands and Roosevelt's legacy, according to a rambling, chaotic biography of Schafer's life and business, written in 2000 by North Dakota's poet laureate, Larry Woiwode. Early in the company's history, Schafer often brought his salesmen west from the company headquarters in Bismarck for rustic retreats. Schafer wanted to inspire his salesmen by showing them how two different sorts of "aristocrats" lived in the West, Woiwode wrote.

Even in the 1950s, Medora was a rough-and-tumble place with little to recommend it to tourists. Schafer began buying old properties and restoring the town, in part because no one else seemed to care or have the money for its upkeep. He told Woiwode that he saw its potential. If four million people visited the Black Hills of South Dakota every year, why not Medora, too?

"He rekindled remembrance of the Marquis and TR, mostly by insisting they were lasting emblems, high water marks, of local culture," Woiwode wrote. "What had brought them to the Badlands was what remained: *The West*."[21]

The Medora Musical.

The Medora Musical is billed as "the Greatest Show in the West," and I wanted the full experience.[22] So I booked us tickets for the Pitchfork Steak Fondue before the show. The buffet cost thirty dollars a person, but it was filling and surprisingly tasty. We piled our plates not just with steak but generous portions of chicken, beans, tomatoes, and garlic bread. Young people in snap-front western shirts served us steaks grilled literally on pitchforks. The house band, The Coal Diggers, played country songs under a pavilion. From our picnic table, we had a view of the Badlands to the south. The wind sent us scrambling for our napkins.

Labor Day had come and gone, and there were few children at the dinner or the show, one of the final performances of the season. Most of the people lined up for grub were dressed casually in jeans and fleece jackets. Many of them were retirees on bus tours that stopped at attractions in the region, including Badlands National Park and Wall Drug in South Dakota.

As the sun went down behind the rock formations, the temperature dropped. We took an open-air escalator down the hill to the mezzanine overlooking the amphitheater. From the mezzanine, I could see the stage for what it was: a false front, just like one of those old western towns with impressive Main Street edifices that hid grubby, utilitarian back rooms. The stage was set at ground level so that trick riders could spur their stunt-trained horses right up on stage. We asked someone to take a picture with our faces peeking out of a plywood cutout of a dancing cowboy and cowgirl. Chris bought us hot chocolate, which came in silver keepsake Medora-themed thermoses. He splashed some bourbon from a flask into his—no booze is sold at the show—and we settled into our seats near the front of the stage, a camp blanket on our laps.

The show lights went up. They cast a red-white-and-blue flag onto the buttes framing the amphitheater. Above it all was the word "Medora," in all caps, just like the Hollywood sign. A rider herded trained elk up the hill, their antlers silhouetted against the sign. The show has always had—and continues to have—three requirements. It must be patriotic and family friendly, and it must be about Theodore Roosevelt. Each show includes a recreation of the 1898 Battle of San Juan Hill, the military campaign in Cuba that made Roosevelt a war hero.

The night we were there, the musical was emceed by a longtime Medora fixture, Bill Sorensen, the former mayor of Bismarck, and his copresenter, the singer Emily Walter. I was, admittedly, a little mesmerized by Walter, a blond performer in her forties who wore sequined strapless column gowns and sparkly cowboy hats in patriotic colors.

Walter told me she got her start as a singer in the U.S. Air Force with the Strategic Air Command band.[23] After her military service, she landed a summer gig in 1993 during graduate school as one of the Burning Hills Singers. Her first time in Medora, she arrived from the east after driving her Saturn sports car eight hours across the flat terrain of western Minnesota and across North Dakota. She, too, stopped at the Painted Canyon exit.

"I just remember being stunned at how gorgeous everything was," she told me. "And then to come into the little Wild West town and see that kind of architecture, that kind of history, and then to be able to perform in an amphitheater, built into a butte with the backdrop of the gorgeous western sky. I mean, it was incredible. After twenty-seven years, I still turn around and see that sun setting or the elk running behind me or the unbelievable colors of the sky. That is the stuff of Medora."

Walter became known for singing Patsy Cline's *Sweet Dreams*. After her first stint in Medora, she worked as a traveling performer, singing on cruise ships, in regional musical theater and in roles portraying Cline. In 2010, the producers of the show invited Walter back to Medora to host the musical. For her role as host, she took on the persona of the "Queen of the West," an elegant cowgirl character suggested by one of the producers of the show. (Roy Rogers's wife Dale Evans was the original television Queen of the West in the 1950s.)

"I pretty well embraced the Queen of the West," Walter said. "I loved dressing up in cowgirl gear, and I love singing the songs. I just like being that person to the guests and all the little girls that have come through Medora."

Little girls were especially drawn to the Queen of the West character, Walter told me. Many approached her on the street in Medora to hug her. Walter thought it was the sparkly tiara on her cowboy hat. I knew it was more complicated than that, but I understood the allure. The Queen of the West was part cabaret singer and part rodeo princess, mixed with a touch of bordello madam and a frontier schoolmarm. White womanhood in the West was confusing. It puzzles me still in my forties, and I *am* a white woman from the West.

We watched as the Burning Hills Singers launched into "Dakota," a ballad that's featured every year. It's from the 1968 Disney musical, *The One and Only, Genuine, Original Family Band*, a limpid film about settlers in Dakota Territory during the 1888 presidential election.[24] The song is by Robert and Richard Sherman, the Oscar-winning brothers who wrote the scores for *Mary Poppins* and *Chitty Chitty Bang Bang*.

In the film, John Davidson plays the lead character Joe Carder, a journalist trying to persuade people to move to the Dakota Territory. As he sings, the camera pans to Carder's love interest Alice Bower, played by Lesley Ann Warren. As he approaches her, Carder sings of virgin fields awaiting a plow.

I cringed, hearing those lines onstage in 2014 from a forgotten, mediocre Disney musical made in 1968. I could almost feel the singers bracing themselves for the line about virgin fields, so that they, too, didn't cringe.

In her review, the *New York Times* critic Renata Adler described *The One and Only, Genuine, Original Family Band* as "about as pepless and fizzled a musical as has ever come out of the Walt Disney Studios" with "several uneasy, convictionless patriotic numbers."[25]

It could have been a modern-day review of the Medora

Musical—notwithstanding Walter's considerable charm and talent, and the musical prowess of the Coal Diggers Band and the Burning Hills Singers. The show we were watching in the outdoor amphitheater in the waning days of the summer felt mired in the decade of its creation. Many of the Burning Hills players are aspiring musicians and dancers, and some are veterans of the Branson-and-Dollywood musical theater circuit. After their summer season contracts end, a few stay on to play Christmas-themed shows in Medora and around North Dakota. I was certain many of the cast members dreamed of Nashville record contracts or Broadway stardom—they said as much in their bios. Yet here they were, in the dusty West, working artists making a living singing and dancing every night for people stuffed with steak after the Pitchfork Fondue. It took grit and perseverance to pursue such a life, Walter said.

"I love it," Walter said. "It's not for the weak of heart, when you've got a show every single night with no day off for four months at a time. And, you know, going on whether you're sick or not, or you've got sore muscles or, something happened personally or whatever, it's something that you just have to do."

Yet I also knew that artists could be powerful agents of change. It seemed such a squandered opportunity to me, to rely on stale storytelling and songs from pepless Disney musicals, especially when there were so many better, more interesting stories to tell about the West. The musical was desperate for an update. All that talent could do better than championing a version of America that never existed, in a town that was only truly alive for three short years.

The next day, we drove through the park so that Chris could see all the bison up close. Later in the park headquarters, we wandered through an exhibit about the threat to night skies from oil and gas development. There's no oil field development inside the park, but in some places, oil and gas rigs are visible on the horizon. The exhibit warned that light pollution at night from the oil fields threatened naked-eye views of the aurora borealis and the Andromeda Galaxy from the park. The park sought to keep nighttime vistas as they were in Roosevelt's time. The exhibit offered up a quote from Roosevelt's own writing about his time in the Badlands: "In the soft springtime the stars were glorious in our eyes each night before we fell asleep."

Roosevelt held despicable beliefs about the Native people who lived in the Dakotas before, during, and after his arrival.[26] But it's also true that if Roosevelt had not spent time in Medora and had not written so much about his time there, the national park that bears his name would not exist. And if the park didn't exist, it wouldn't have served as a bulwark against oil and gas development on every available inch of western North Dakota.

We stopped inside the park headquarters and the gift shop. Chris was enamored with a stuffed baby bighorn, like the rams we'd seen in the bachelor herd on our South Dakota hike. I could picture the soft toy in a crib, along with the baby bald eagle and the prairie dog puppet also for sale. We both laughed at *Who Pooped in the Park?* a guide to the animal scat and tracks of America's national parks. I flipped through the pages of *The Girl Who Loved Wild Horses*, the sort of book I would have begged my mother to buy for me when I was a girl.

I couldn't buy the book or the plushie. Not until we really knew for sure we were going to have a baby. I also decided not to add Medora to the list of possible baby names. Medora, I learned, spent her later years defending the reputation of her husband, who became consumed with anti-Semitism after he returned to France. He traveled to North Africa in 1896, where he was killed just a few days short of his thirty-eighth birthday. The marquis was there as part of a grandiose and largely self-conceived scheme to conquer all of Africa for France.[27]

The next day Chris and I drove toward the Bismarck airport, listening once again to *Undaunted Courage*. We had reached the point in the book where Lewis and Clark have traveled over the Bitterroots from the Pacific Northwest. They're about to venture off onto separate journeys, before reuniting for the trip back down the Missouri to St. Louis. Clark was exploring along the Yellowstone; Lewis, the Marias River.

It was fitting literary punctuation to our journey. Chris needed to return home for work. I was continuing on in North Dakota to do more research about Anna and some reporting for a few freelance pieces. We planned to reunite at home in Washington, DC, in a few weeks.

The journey had been a temporary reprieve from the world of infertility treatments, yet a baby was never far from my mind. My doctor told me it was possible I could get pregnant without intervention, and I would know soon whether the rock back in Illinois really worked.

A few days later, all alone, I peed in a cup in the bathroom of a motel just off the interstate in Bismarck, across the street from a drive-through Starbucks. I dipped a pregnancy test strip into the cup, and I set my phone timer for five minutes.

It was negative. The mystery baby-making rock of Liberty, Illinois, was a cock-shaped glaciated granite boulder, nothing more.

14

THE BUST

December 2014

North Dakota crude: *$50.51 per barrel*

ANOTHER MANILA ENVELOPE FROM AN OIL COMPANY ARRIVED IN MY father's mailbox in Oregon. This time, Cornerstone wanted to pay $4,000 to extend my father's existing lease for another two years.

The offer and the amount of money surprised me. Oil prices were steadily sliding, and most analysts predicted that the industry faced a serious slump, caused in part by oversupply in North Dakota and other new oil plays in North America. But the lease renewal also tracked with what I had learned in Colorado about smaller, independent producers. Slumps were when investors with cash put their money into bargains they believed would pay off in better times. Downturns were when they went in search of opportunities ignored by the major oil companies, outside of the frenzy.

I sought a lawyer in North Dakota to make sure all the paperwork was in order for my father, even if it was unlikely Cornerstone or any oil company would ever drill on Anna's land.

The first lawyer I called was a thirty dollar referral from the North

Dakota Bar association. I told him I wanted someone with "a spirit of adventure," which was a mistake. To him it was business, not participation in a creative endeavor. Nonetheless, the lawyer gave me the lay of the land: In North Dakota, attorneys with experience in oil and gas and mineral rights were overwhelmed. Many in Minot were still busy with land use work following the floods several years earlier. They had nothing remaining but a "spirit of fatigue," the lawyer told me. He gave me about fifteen minutes of a promised thirty-minute consultation, but declined to represent me. My business was unlikely to be worth his trouble, and I certainly wasn't going to lead to lucrative ongoing billable hours.

The next lawyer I called was a solo practitioner who didn't mind being quoted, but he had a scheduling conflict and canceled on me. And the third lawyer, part of a big firm with a long history in North Dakota and Minnesota, ran it past his partners. They wished me well on what they called my "venture," but they declined, too. Almost all of the lawyers were wary of seeing their advice in print.

Their reluctance to take me on as a client reminded me of Anna's legal troubles when she sought to abandon her first homestead claim for another. "Such defects in procedure as there may have been in the past are not her fault, but the fault of her former attorney," her new attorney wrote in the paperwork she filed with the General Land Office. Finding a good legal advocate was apparently a multigenerational family challenge.

Finally, I got in touch with the only lawyer in North Dakota seemingly willing to accept a journalist as a client: Derrick Braaten. Based in Bismarck, Braaten went to law school so he could represent ranchers and farmers, but he ended up becoming an expert in mineral rights because of the oil boom.[1] Many of his clients were landowners who saw adverse impacts to their property from the boom, or who had conflicts with pipeline companies about construction or even oil spills. Braaten was also politically independent, which made him more at ease being a contrarian in Republican-dominated North Dakota.

Our conversations were brief. I had a tiny budget carved out of some of the money my father got from the new lease, but Braaten charged $250 an hour. I gave him all the paperwork I had. I also shared what I learned from the geologists in Colorado, who didn't think anyone would ever drill on Anna's land. Braaten wasn't optimistic, either.

The warning signs were everywhere. The U.S. in 2012 surpassed Saudi Arabia to become the world's largest oil producer.[2] In June 2014, North Dakota marked the production milestone of a million barrels of oil per day—celebrated fittingly in Tioga, the site in 1951 of the first commercial oil well in the state.[3] By late 2014, the U.S. was producing 8.7 million barrels of oil each day, a million more per day than a year earlier. The national glut derived from not only the Bakken, but from new oil fields in Colorado, New Mexico, Texas, and Wyoming and beyond. "Suddenly the world is awash with oil," the *Financial Times* warned mid-December 2014. "The scale of the current oil shock is difficult to exaggerate."[4]

Demand was also weak globally. OPEC, the international oil cartel that influences worldwide prices by regulating supply, declined to cut production among its thirteen oil-producing member nations. The glut meant lower worldwide prices. In October, the *New York Times* warned that "slowing American oil production is like slowing a freight train moving at high speed."[5] It was a poor choice of metaphor considering the industry's recent history of oil trains derailing in fiery explosions.

The boom helped the United States reduce its dependence on imported energy, the Council on Foreign Relations said in its own forecast. But it had a warning for states like North Dakota. "Some of

the states providing new energy resources are becoming less economically diversified and more economically vulnerable to energy price
declines."[6]

Everyone in North Dakota was nervous. "If this is really only a
slow period, tell us," pleaded the editors of the *Williston Herald*, as the
price of oil plunged in December. "If the state has real fears the bubble
is bursting—again—tell us. If nobody has any clue as to what the future
holds, and what is happening right now, tell us."[7]

The drop in oil prices was reported on national newscasts in the
final months of 2014 as a matter-of-fact economic indicator. The prices
were uniformly portrayed as being good for consumers but bad for
American oil producers and the economies of booming places like
North Dakota.[8] "First, some good news," an NPR announcer said in
early November. "Oil prices fell again today. And that's great for all
of us who've been enjoying lower gas prices. But it's not good news
for U.S. oil producers. They've become important players in the U.S.
economy and in the global oil market. And for them falling oil prices
are a threat."[9]

Left out of most discussions in most of the mainstream media was
how the U.S. fracking boom, with its unchecked methane flares, contributed to the greenhouse gas emissions responsible for climate change.
There was little effort to tie oil prices and what was happening on the
ground in places like North Dakota to the worldwide climate crisis.
"The more we delay, the more we will pay," warned Ban Ki-moon, the
United Nations Secretary-General, at a December 2014 climate conference in Peru to prepare for the Paris Climate Accords the next year.[10]

His alarm went unheeded. No one could say no to the boom, especially not in North Dakota where production taxes were burning a
hole in the pockets of state government officials. Feeling flush, North
Dakota Governor Jack Dalrymple proposed a state budget that gave

local communities more of the money collected from oil and gas production taxes. Dalrymple also asked the state legislature to approve $873 million in one-time spending to help oil patch communities manage growth. Of that, $450 million went to the state Department of Transportation to transform two-lane roads into four-lane highways, and to build bypasses around towns to manage oil field traffic.[11]

The money and the reconfigured oil tax formulas were exactly what people wanted in western North Dakota. Even so, an editorial in the *Williston Herald* just before the holidays warned the governor to be realistic with his budget projections. The governor's budget was based on projections that set oil prices at seventy-four dollars per barrel. It was a number that was beginning to look as though it, too, would forever be in the rearview mirror.

"We don't want to end up like Clark Griswold and buy the pool before the Christmas bonus comes through," the *Herald* wrote.[12] Reading the editorial from afar, I laughed out loud at the lack of self-awareness. Had they forgotten in Williston that they blew their bonus on a $76 million pool, one with a fancy waterslide? They were, pun intended, already under water. Totally out of their depth.

I kept track of oil prices and the effects of the slump, unsure how it would affect the mineral rights that were Anna's legacy. "I've been watching the price of oil and wondering whether the company is going to follow through on the leases," I told Braaten in an email. It felt like something was over, something that had barely begun for me.

For Christmas, Chris and I agreed on a one hundred dollar spending cap on gifts for each other. Our insurance covered the basics of IVF, but we were responsible for thousands of dollars in out-of-pocket costs for fertility drugs and copayments. It left little wiggle room in our budget for other indulgences.

I ordered Chris a Kaffir lime tree, which I hid behind the shower curtain in the tub in our rarely used basement bathroom. I packaged the gift with a Thai cookbook featuring recipes with lime leaves. Then, I attached a ribbon to the book. Chris had to follow the ribbon down to the basement to find the lime tree on Christmas morning. We needed the playful distraction. In a few days, we would learn whether our first IVF cycle was a success.

Chris gave me a pair of tan, sheepskin-lined Ugg boots. The boots were just barely within our one hundred dollar limit. They were a cozy but not especially fashion-forward gift. Uggs were a uniform of East Coast college girls half my age, who wore them with black Lululemon leggings, North Face fleeces, and their hair tied in topknots.

I didn't care, I loved my cozy Uggs. They meant Chris had been listening when I told him what my acupuncturist told me: Chinese medicine suggests that women who are trying to get pregnant keep their feet warm. My acupuncturist also had me place a buckwheat-and-lavender heating pad under the small of my back each night before falling asleep. The heating pad was supposed to bring warmth and vitality to my uterus. Chris, who ground coffee beans each night before bed and set the timer on the coffee maker, added the heating pad to his nightly ritual. He warmed it for me in the microwave while he ground the beans, and then carried it upstairs to me in bed.

There were suddenly many such new rituals in our lives. They all came after I read *It Starts With the Egg*, a book popular among older women trying to have babies. The book was particularly wary of the effect on egg quality of BPA, a compound used to make plastic harder, and one in wide use in consumer goods. BPA was especially dangerous when heated, the book warned.[13] So Chris researched and bought a replacement coffee maker after I read that BPA was in the plastic components of most coffee makers. Reluctantly, we donated our souvenir Medora coffee cups with their cheap plastic lids.

Meanwhile, I limited myself to just one cup of coffee per day. We added maca powder and Brazil nuts to our daily smoothies—superfoods that were supposed to boost sperm count and egg quality. Deep down, I doubted any of it mattered. The superfoods and the heating pad and the new coffee maker and all the yoga and acupuncture were like the Mystery Baby-Making Boulder. They made us feel as though we had some sort of control over the outcome.

To keep ourselves occupied while we awaited results, we went to the movies. We saw anything to keep our mind off our wait, including *Into the Woods*, the cinematic version of the Stephen Sondheim musical.[14] I last saw the play—or at least part of it—on Broadway the summer I turned fourteen. Just a few minutes into the production, though, I had been overcome with food poisoning. I barfed all over myself as I ran out of the theater onto the New York City streets. My mother, who waited in line at the World Trade Center for discounted same-day tickets so we could see the musical, trailed worriedly behind me. She got me back to the hotel, and then returned to the theater. She really wanted to see the show, and I was old enough to take care of myself.

Decades later, I nearly exited a theater again, this time in Washington, DC. As a play, *Into the Woods* has many layers, including the allegorical story of the AIDS crisis of the 1980s. The on-screen narrative of 2014 was driven by the story of a childless baker and his wife, who were cursed with infertility by a once-beautiful witch. To lift the curse, they had to venture out into the woods the first three days before the rise of a blue moon to find ingredients to restore the witch's beauty: a milk-white cow, hair as yellow as corn, a blood-red cape, and a slipper of gold.

Maca powder and Brazil nuts, I thought to myself as I watched the film. A BPA-free coffee maker and a lavender heating pad. My Ugg boots.

We squirmed in our seats.

"Do you want to leave?" Chris whispered to me.

"No," I said, unwilling to miss the story yet again. "We might as well find out what happens."

On New Year's Eve, I drove across town to the fertility clinic for a pregnancy test.

It was negative.

That night, friends joined us for homemade pizza, champagne, and a raucous round of Cards Against Humanity. I drank only a glass, but the next morning, I felt slammed by the hormonal crash of a failed IVF cycle. It made me feel as though I had a mean, red hangover, the kind where every past decision haunts you and the future looks hopeless.

I snapped angrily at Chris, who forgot to turn on the BPA-free coffee maker before going to bed. As I cleaned up from the night before, I broke three champagne flutes. My whole body felt just as fragile, my inner state just as jagged. The year in review played in my mind. Nothing was going right. I wanted to be pregnant. I wanted Anna's mineral rights to mean something. I felt stuck in a repeating pattern of disappointment, one that I had to revisit every month. It had been just three months, but our happy road trip through the Badlands seemed so long ago.

The dirty dishes overwhelmed me. I fled the house, in search of an open coffee shop. I ended up at a favorite spot in Rock Creek Park, near where Chris and I shared a first kiss. I sat alone on a rock, bundled up in my puffy down coat. No one else was there. I listened to the sound of water rushing over rocks and felt the cold rock beneath my bottom. I drank the hot coffee. The first day of the year was sunny and relatively mild. The sun struck the surface of the fast-moving creek in such a way that it cast rippling, dappled reflections on the trunks of the trees lining the bank of the creek. Beauty, replicating itself in light, changing by the second. Too fluid to be fixed.

Even though I knew exactly what I yearned for, I decided not to make any resolutions. Rigid expectations would set me up for failure and disappointment, especially when so much was out of my control.

A few days later, Derrick Braaten called with an update from North Dakota.[15] He needed to do much more extensive legal work to establish clear title to the mineral rights in order to ever receive royalty payments from the oil company. It meant re-opening probate on the wills of my long-dead grandparents, drawing up an affidavit to explain why my mother didn't have a will, and searching the courthouse files for official, notarized records. He estimated it would cost between $10,000 and $15,000.

Braaten didn't have to say it. The legal work wasn't worth the expense, even if I had the money to spend. North Dakota crude was $37.52 per barrel the day of our phone conversation, a full sixty dollars per barrel less than when I first flew to North Dakota in search of Anna. The lease money might keep coming. But even if an oil company drilled on Anna's land, my family's share of the oil pumped out of the earth was so small it would take years for the royalties to cover the legal costs of establishing our rights to them.

Braaten wished me well. I wrote his firm a $641.86 check for the work.

Just a few weeks later, the Federal Reserve Bank in Minneapolis had a dismal forecast for the Bakken: "The drop in oil prices has slowed drilling activity, and energy companies have announced that they will scale back future drilling operations." The Federal Reserve suggested an even bleaker outlook for prospects on the margins of the oil patch, like ours. "These estimates suggest that operators will suspend most exploratory drilling on the periphery and focus on drilling in the core area of the Bakken, where costs are lowest."[16]

The boom was about to go bust. Even oil tycoon Hamm took a hit. His personal fortune was worth $18.7 billion in August when *Forbes* compiled its list of 400 richest Americans. By the end of 2014, oil prices had crashed, and he owed his ex-wife nearly $1 billion as part of their divorce. *Forbes* reported that Hamm's net worth, tied up in shares of Continental, dropped to $8.9 billion. Still a multibillionaire, Hamm nonetheless appealed the divorce settlement, saying he needed to borrow money to pay his wife the cash she was owed in the split.[17]

The whispers of wealth on the Great Plains were beginning to seem like a ghostly croak of desperation by an industry too greedy and too strung out on its own supply to ever consider the long-term consequences of its actions.

15

THE DOLLHOUSE

September 2015

North Dakota crude: *$37.78 per barrel*

We rolled open the garage doors to the late summer sun of western Oregon and turned on the radio. One side of my father's two-car garage served as a packing room to send out the finished jewelry boxes he made in the workshop behind the house. The other was filled with a large work table and shelving, lined with dozens of Rubbermaid bins and cardboard boxes. Some of the boxes were unopened from when he and my mother moved into the house more than a decade earlier.

The boxes were dusty and covered with cobwebs. Dislodging all the dust made us feel itchy and as though we were on the verge of sneezing all afternoon. Some of the boxes were warped with water damage. The well water at my father's house was acidic, and a few years earlier, it had worn through the copper pipes and flooded the house.

My father owned a lifetime's worth of the stuff all Americans accumulate, made worse by being a professional woodworker who liked tools. It was daunting enough inside the house and the garage; the contents of my father's professional woodworking workshop required the services of an auctioneer.

Chris helped by taking boxes off the shelf and moving discarded items to the trash pile. There wasn't much else he could do. It was mostly up to me to make the hard decisions my father couldn't about what to keep and what to throw away before he sold the house and moved in with his girlfriend.

Some decisions were easy. "Dad," I asked, "do we really need three copies of your master's thesis on adhesives?"

"Uh, do we even need one?" Chris joked.

I wiped the grit from my hands on my shorts before untucking the cardboard flaps on a box of photos. Out wafted a powerful whiff of disintegrating negatives, old leather, and dust. Inside, I found a slide projector, three carousels of slides, a rolled up tube of papers, and several old books, including a copy of Shakespeare's *Twelfth Night* with my grandfather's name inside the cover. The tube held his high school diploma and a grade book showing he attended additional classes in Fargo. One bible was in Norwegian. My great-grandfather's name, Andrew, was inscribed in it, but there were no other names.

There was also a small, palm-sized book, no more than half an inch thick and bound in faded black cloth with a pleasant, tactile feel underneath my fingers. It was a Norwegian church songbook, entitled *Sange paa Livets Dei*. It appeared to be a gift to Anna, before her marriage. Someone had inked her name inside the cover, in a neat cursive: Miss A. Sletvold. Underneath her name it read "Erindring fra O. Landmark." Roughly translated, it meant "a token from O. Landmark."

There were nearly two dozen blank pages at the back of the book. Someone wrote on three of the pages, in pencil. The writing on the first page was also in Norwegian, but in a different hand than the one that wrote the inscription to Anna at the front of the book. "Edwin H. Haraseth is born, 29 Dec. 1906."

The next page was blank, except for a date, also written in pencil: April 5th, 1907. I flipped to the next page, where there was one final notation: Dec. 13, 1905.

I felt chills. I knew all of these dates well. The first was my grandfather's birth. The second was the date Anna entered the asylum. The third was the day she and Andrew married. All three added up to the meager timeline of a woman who, in the space of eighteen months, went from pretty young bride and homesteader to mother and then to inmate.

Like Anna, I also understood how just eighteen months could change your life. It seemed all I had done for the past year or so was try to get pregnant. "Tick tock, biological clock," I wrote to Chris, in yet another calendar invite for yet another appointment at the fertility clinic that year.

I got pregnant, once. The day of my pregnancy test, something was wrong with the calibration at the clinic's blood-testing machines. The nurse who called to share the news couldn't officially tell me I was pregnant. She asked me to take a home pregnancy test and email her the results. I laughed out loud in a park near my home, where I'd stopped to take the call while I was on a run. Instead of the happiness I hoped to experience with such news, the nurse left me with uncertainty. It was not a good omen.

By week eight, things got worse. Chris and I drove to the doctor for our third ultrasound to see whether "the pregnancy" was progressing. This was how our doctor described it. Not "the baby," or even "the embryo" or "the fetus," but "the pregnancy," an awkward phrase necessitating the passive voice.

I was on the exam table, undressed from the waist down. Chris stood next to me, on the left, his right hand grabbing my left. The screen was on the other side of my body. Our doctor was positioned between my legs. Chris leaned across my belly, toward the ultrasound screen. I could feel the tension and anxiety in Chris's hand. It was almost painful, the way it made the hard stone of my wedding ring press into the side of my pinkie finger. He wasn't squeezing my hand to reassure me; he was anxious because he couldn't see the screen. He was desperate for visual information.

"I'm blind as a bat," Chris reminded our doctor, who enlarged the image and put a tiny cursor on it. It was an unusual request from Chris, who hated calling attention to his limited vision. I don't think the doctor understood. I pointed to the screen for Chris; the cursor was way too small for him to see. At that moment, more than I ever had, I wanted Chris to be able to see well.

But there was nothing to see in the white blob. No heartbeat, no fluttering on the screen like there had been at the last appointment. "The pregnancy" was over, or it would be once I took misoprostol that afternoon to induce a miscarriage.

My doctor touched me lightly on the shin. His compassionate touch surprised me. He was very kind, but as a male health care provider, he was understandably cautious about touching female patients. The shin was both a weird and surprisingly reassuring place to make contact.

"Take your time. I understand that this must be devastating," he told us.

It wasn't devastating. It was something worse, I thought, as our doctor left the exam room and I got dressed. It was exhausting. I knew that if I did another IVF cycle it would eat up at least another three months of my life. I would have to recover for a month from the miscarriage. I'd be on birth control pills for two weeks. Injections daily for another two weeks or so. Monitoring ultrasounds every other day or so. The time driving to and from appointments. Weekly acupuncture. An invasive egg retrieval, requiring anesthesia and a day away from my work. The bloating and fatigue from vaginal progesterone supplements. So much yoga. Weeks of waiting on an uncertain outcome, yet again.

Chris and I walked to the elevator. Some cruel baby had dropped a pacifier in the carpeted vestibule in front of the elevator. I laughed out loud, incredulous at the heavy-handedness of the symbol. It was so maudlin, so obvious.

"Is that really there?" I said, pointing it out to Chris. "Really?"

Chris and I left the building and walked toward my car, the hoods up on our raincoats. The light, warm drizzle and gray skies matched our mood. A beautiful spring day would have been an affront to what had just happened, I told Chris. Like that pacifier.

———

The dollhouse.

Four months later in Oregon as we sorted through the past, my father brought out from his workshop the dollhouse he made for me when I was four. He had been storing it on a high shelf in his woodworking shop, so it was covered with sawdust. The dollhouse had two stories and an attic room. There was a chimney on one side. On the other side, stairs ran up the yellow exterior to a walnut door, hinged with the same fixtures my dad used on the jewelry boxes he sold in galleries. Inside, wallpaper peeled away in strips. The dollhouse needed major cosmetic renovations for the next generation, but like all the things my father built, it was of solid construction.

I wanted the dollhouse, desperately. But I couldn't bear for my father to pack it in the crate. Until then, I hadn't told my father about our specific failures, including the miscarriage. It wasn't as though I was hiding anything from my father by not telling him. He knew we were trying to have a baby, but I just couldn't bring myself to ever share the sad details. My father went through enough bad news as my mother's health declined. All I wanted to do was to share good news with him, when I finally had good news to share. It meant I hadn't been able to share much at all with him. There in my father's dusty garage, with the dollhouse on the work table, I explained: Good news was unlikely.

"I can't take it with me," I told him.

My father was quiet. He didn't know what to say. All he did was nod as I told him why we couldn't pack the dollhouse in the crate. He gave me a hug, one of his side-body squeezes that could leave you breathless. I could tell he was sad for us, and disappointed. I knew he would have enjoyed being asked to renovate for a grandchild the dollhouse he made for his first daughter.

I went to the bathroom to wash my face and forearms of all the dust. There were tracks from my tears in the dirt on my cheeks. I laughed at the mirror. It was a real phrase, tear-streaked. It was a thing that really did happen.

While I was gone, my father carried the dollhouse away. It went in the "donate" pile, not the trash. Someone else's child would get it. Someone who saw a windfall in a fixer upper.

The bulk of the remaining boxes in the garage were my mother's, and she was both a packrat and archivist. Inside one box was a white ceramic planter with a baby elephant on the front. My mother taped a note on the back of it, documenting exactly when she received it: "Vase from flowers sent to Steph and Nan in hospital from Bill." It was so important to her that it called for annotation, yet neither my sister Steph nor I knew why our mother felt as though she needed to keep a vase our father sent her when Steph was born.

There was a whole box of collectible blinking-eye Nancy storybook dolls, which people gave my mother in the 1950s because she shared their name. As children, the Nancy dolls were off-limits to me and my sister. They were, allegedly, valuable collector items. So, they sat unused in a box on the garage shelf for decades, taking up space. Then there were baby clothes, both mine and my sister's. Scraps of my mother's wedding dress fabric, mildewed by age and inattention.

Out of a sense of due diligence, I texted my sister: *Should I keep the fabric for your daughters? No*, Steph wrote back, with the ruthlessness of someone not actually handling the stuff. *Toss all that shit.*

We both felt the twinge of future regret, but neither one of us wanted much more than the photos, some furniture and some artwork. Moving was made much worse by having to sort through the past and its objects. All this stuff had a cost. I could see the psychic weight on my dad of the burden of all these dusty boxes. Nostalgia could immobilize you with indecision for days.

I paged through the scrapbooks of my mother's high school and college years, reluctant to toss them and equally reluctant to pack them into tubs I would never open. My mother's earliest scrapbook was from 1945. In it were notes and cards documenting all the baby gifts my grandmother received at a shower before my mother was born. I flipped through them, and then, before I could change my mind, threw them on the ever-growing garbage pile my father planned to take to the dump.

The notes and postcards and prom programs and newspaper clippings about forgotten 1950s celebrities weren't important. What I took from them was: my mother was loved and cherished even before she arrived, and when she got here, she loved documenting the world around her. These objects and scrapbooks meant something, if only to define her past and establish the person she had become. Now she was gone, leaving her daughter to make what meaning of it she could.

I, too, had several boxes to go through. There were childhood books, including the one my mother used to explain how babies were made. I found papers I wrote in high school history class, a sheaf of slides from my first photography class, and other forgotten detritus of childhood. *John Denver and the Muppets: A Christmas Together.* A unicorn refrigerator magnet. A rainbow charm bracelet. An orange ball cap from my brief time on the middle school all-star softball team.

Buried in one of my boxes, I found a note my mother wrote me. Until her strokes and heart attacks, my mother wrote with a clear, friendly cursive. I loved the way she wrote the "E" in my first name. It was round, with an old-fashioned curlicue at the top. I liked to think that every time she wrote my name or my sister's, it was an expression of affection. Her handwriting was so distinctive that, at age seven, I began doubting the existence of Santa Claus because I recognized my mother's script on the "to" and "from" tags on our Christmas presents.

The note I found was a five-item list of instructions for my day, written early in the spring of my last year of high school. Some of the tasks on the list were simple: "Pick up the camera before school," my mother wrote. Others were more complex maneuvers that required focus, charm, and a deliberate approach: "Call about a summer internship," she reminded me. One was a warning not to lock my keys in my car again. And the final task? Have some fun: "Have a great day, and a great, fantastic time at the game tonite."

At first, I wondered why I saved the note. It seemed on its face like a straightforward to-do list from a mother to her unfocused, seventeen-year-old daydreamer of a daughter, a girl who got overwhelmed by the adult tasks she had to finish before she could go to the basketball game. I read it again, looking for clues. My mother signed it not as "Mom," but as "Nan," her name. It was a loving signal: as one adult to another, I am helping you understand how to manage what seems overwhelming. The subtext: You can do this. Here is how: Make a list of your tasks, check one thing off at a time. Then go have fun. Love You! Nan.

A wave of shame washed over me as I read my mother's words. It had been a year since my last visit to North Dakota. Almost without realizing it, I had allowed Anna's story to slip away from me. A story that seemed so vital to tell seemed to recede into the background, overcome by the sense of urgency to get pregnant before it was too late. The insistence of it enveloped me in a disorienting fog that only occasionally cleared, and when it did, I was dismayed by how quickly I was willing to set aside a pursuit that had been—that was—so important to me. I was writing, but not about Anna. Half of my income that year came from teaching yoga, which I enjoyed, but I hesitated to call it a career. It was a low-stress job I supplemented with freelance writing—even as I put the bulk of my energy into having a baby while I still could.

If one measure of being a good daughter was living up to your potential, I was failing at it. My father had said as much when he asked why I was teaching yoga and told me he missed reading what I wrote. I had all but abandoned Anna's story in pursuit of a child of my own that year. What had happened to my spirit of adventure?

It was impossible to know who made the notes in Anna's songbook. But I was the only person alive who understood their significance. I was the only one who could tell the story of those dates. No one else. It was a sign from the past, urging me back into action.

I tucked Anna's songbook in my purse. Everything else went into five Rubbermaid tubs crammed with photos, slides, and the slide projector. The tubs would keep out dust, and they could be packed into the empty space in the furniture crate. I could store them in my own basement until I had time to sort through them myself.

It nagged at me, the unfinished business of both the past and now the future, but it couldn't be helped. A more detailed sort awaited me.

I tucked my mother's note in my purse, too, in the same pocket as Anna's songbook. At the very least, I owed my mother an account of what I found when I went in search of the story of Anna's life. To do that, I needed to find a way to get back to North Dakota.

Part III

16

HYDROCARBONS

June 2016

North Dakota crude: *$42.80 per barrel*

JULIE LEFEVER FROWNED AT WHAT SHE SAW ON HER COMPUTER IN HER office on the second floor of a science building at the University of North Dakota.[1] "Not much success up in this area at all, with the Bakken," she told me.

LeFever was the oil oracle of North Dakota, a geologist whose work predicted the promise of the Bakken formation. She was also the geologist in charge of the core sample library in Grand Forks, on the eastern side of North Dakota about as far as you could get in the state from the actual oil fields. The 39,000-square-foot warehouse stored 85 miles of rock cuttings unearthed by geologists searching for oil.

Even after all I had learned, I wanted an official pronouncement: Would anyone ever drill on Anna's land? I also wanted to see whether LeFever had any core samples from the area where Anna once staked a claim.

In person, LeFever was soft-spoken, with the demeanor of an experienced, patient professor who enjoys explaining difficult concepts to her students. Like the geologists I interviewed two years earlier at the U.S. Geological Survey (USGS) in Denver, LeFever seemed to take pleasure in sharing what she knew with me. Like them, LeFever was a woman in a field dominated by men.

At her computer, Julie zeroed in on wells in Burke County, near Anna's original homestead. Most of the wells we looked at on her screen were older efforts from earlier oil booms, drilled into the Madison formation using more conventional drilling techniques. The few Bakken wells drilled nearby were by no means gushers—the first well that LeFever pulled up on her computer was a dud that drew up just eighty barrels of oil per day.

"That's not much," I said, trying to do some quick math in my head. According to the lease, my family got one-eighth of each barrel. But we held only 40 percent of the acreage, which was split by four heirs. Back when oil was selling for prices closer to $100 per barrel, it would have meant about 2 cents per barrel. That added up to about $500 per year. Oil prices were currently around forty dollars per barrel. So, if a well only drew eighty barrels per day, it added up to less than $200 per year in royalties, according to my sloppy math.

"No, that's not much," LeFever said. "A lot of these wells cost between $8 million and $13 million to drill. You're not going to get your funding back very quickly."

A blockbuster Bakken oil well produces 4,500 barrels of oil per day, LeFever told me. The three Bakken wells we could see on-screen in the vicinity of Anna's place were producing only 80 to 193 barrels per day. "That's not economic, as far as the Bakken's concerned," she said.

It was still possible for someone to hit paydirt on Anna's land, she said, but unlikely. Because so few wells had been drilled near Anna's old homestead, oil companies had scant evidence of what the geology looked like beneath the surface. And because what little information they did have wasn't promising, few companies wanted to take risks in the area. They would either need more information to drill, or they'd do it after they had exhausted other more promising sites in the state, if oil prices were high enough.

"It doesn't mean there's not something way down there," LeFever said, "but we don't have anything that would say what's going on with that area. The deeper you go, the higher the risk. And it's convincing a bank or the investors to take that risk."

LeFever called over to the core sample library to ask a graduate student to pull samples from a dry well drilled near Anna's place in 1965, before fracking and horizontal drilling. We walked over to the library to see what the layers of earth looked like beneath the area near the homestead.

Julie LeFever, University of North Dakota, Grand Forks.

Inside, sturdy orange steel shelves were lined to the ceiling with flat cardboard boxes stacked on top of each other, all identified by number. It looked much like the warehouse part of an Ikea store, the section where customers pick up flat-packed, ready-to-assemble furniture. In this warehouse, though, the long, narrow boxes contained rock samples dating back to some of the earliest wells in North Dakota.[2]

Construction was nearly complete on an expansion of the library, paid for fittingly with $13.6 million in money collected by the state in oil taxes.[3] It was named after Wilson M. Laird, a name I recognized from Bill Shemorry's account of the night in 1951 when the first producing oil well in North Dakota came online. Laird was a state geologist and among the earliest believers in North Dakota's promise as an energy state. To watch that first oil well come in, Laird took the train across the state from the University of North Dakota in Grand Forks. And it was Laird who wrote the first state rules for oil companies drilling in North Dakota, including a requirement that they share their core samples with the library.[4]

The core sample library serves both science and commerce, but it is an unabashedly pro-drilling institution. Geologists have a good understanding of what is beneath the earth because of the library. Oil companies have equal access to the information, including the core samples drilled by their competitors.

"Geologists like data," LeFever told me. "They also like rocks."

A graduate student unloaded the cardboard boxes on a broad plywood table. LeFever opened them up. She sprayed the dark gray rocks with water, releasing a smell like fresh rain on a gravel road. We could see coral and brachiopod fossils, evidence that the section of the earth we were looking at was once an open marine environment with a muddy seabed. Then she sprayed the rocks with a 10 percent solution of hydrochloric acid.

"Basically, it either fizzes, it slightly fizzes, or it doesn't fizz," LeFever said. The samples fizzed. That meant there were hydrocarbons there—oil or natural gas.

"You can actually see the hydrocarbons or the organics in this," she told me. "On the top of the bubble, there's kind of an iridescent shimmer. Like you would see if you spilled oil on a wet piece of concrete or something like that. And that would tell you that there's probably hydrocarbons in there."

The formation was too tight, LeFever said, which meant it wasn't very permeable. In 1965 when the well was drilled, there was no technical way to get the hydrocarbons out of the rock. Even now, it would take a creative approach to frack such hard rock, she said. Most producers would forgo such a difficult formation if they had easier, less expensive prospects. "Oil companies have kind of tunnel vision when they're drilling something that they know works," she said. "That's what they target."

To study geology even casually reminds us that humans represent the teeny tiniest speck on a 4.5 billion-year-old planetary timeline. It can be unsettling to contemplate the vastness of Earth's timeline and our human existence in it. Yet I found it grounding to see, feel, and smell the core samples, as well as to hear the acid fizz at the presence of hydrocarbons. If it weren't a completely inappropriate act, I might have licked the rocks, just to incorporate all of my senses into understanding something so vast.

Instead, I stood there at the table with LeFever, imagining the heat at the earth's core and the weight of time and pressure it took to transform a lake bottom into an oil field. Modern drills and high-pressure chemical cocktails could blast through all that time and pressure in mere days, just so we could drive unhindered down the highway, oblivious to the formations beneath us and the invisible emissions above us.

LeFever died just six months after our meeting.[5] I thought she might be ill—she was wearing a wig during the interview—but I didn't know her well enough to ask her about her health. She was sixty-three, the same age as my mother at her death. I read LeFever's obituary in the *Grand Forks Herald*, full of regret I wouldn't be able to follow up with more questions. I knew nothing about LeFever's personal life, other than that she collaborated frequently with her husband, Richard. He, too, had been a geologist at the University of North Dakota. There were two adult daughters mentioned in LeFever's obituary, and I felt a connection with them. Like her daughters, I also felt the grief of losing a mother too soon.

Her daughters didn't know this, but their mother nurtured me, too. Thanks to LeFever's tutorial, I had seen deep beneath the earth of Anna's land. The rocks she showed me confirmed something I'd known for several years, ever since my trip to Colorado: it was unlikely the mineral rights would ever amount to riches for my family.

LeFever was famous in her field, so well-known that I was puzzled by a headline I read in a University of North Dakota publication after her death. The article was about LeFever's posthumous award from the American Association of Petroleum Geologists. It was titled "Largely unknown legacy of Julie LeFever." The headline seemed unfair, sexist even in its assumptions. An accomplished academic, LeFever published more than one hundred papers, including a 1992 report she and a coauthor wrote predicting the potential of the Bakken formation two decades before the boom peaked. She was no splashy oil tycoon like Hamm, but her academic record was hardly an unknown legacy.[6]

Her obituary also described her as a generous researcher: "It was her readiness to share her wealth of knowledge, much of it unpublished, with others which set her apart from most researchers. Countless industry geologists, professors, and graduate students often benefited from Julie's spontaneous, one-on-one core workshops."[7]

"Julie knew where the oil was," her colleague, Kent Hollands, told the university publication. "She looked at nearly every Bakken core, logged it, and put that information together. She knew it was there before the technology existed to extract it."[8]

LeFever was not unknown, not at all. Her legacy of intellectual curiosity created the conditions for an oil boom that brought billions of dollars and thousands of jobs and people to North Dakota, including me. When the state expanded the core library, the university named one of its labs for her.

I wanted to believe it was possible to appreciate the contributions of a groundbreaking woman of science such as Julie LeFever, even as I understood that research like hers most likely hastened the very conditions I returned to North Dakota to write about.

———

It was my first time in North Dakota during the early days of summer. After nearly two years away, I was finally back, this time to report stories about climate change for my new employer, *E&E News*. I had just spent a week on a fellowship with a group of other journalists invited to the state to learn about natural resource issues facing the Great Plains. As I drove, the rolling hills of the prairie pothole region were green with early summer rain. Wild roses bloomed along the side of the road. Prairie smoke wildflowers bloomed a dusky pink. The skies were big and blue with the potential of the long days surrounding the summer solstice. The beauty of the prairie stretched out before me, the promise of the open road a balm after such a fallow period in my personal and professional life.

The climate peril at hand was invisible as I drove west across the top of the state. No state other than Alaska was warming faster than North Dakota, the 2014 National Climate Assessment warned.[9] (When the federal government released an update to its 2014 report in 2018, scientists had another uncomfortable observation about the region: "The energy sector is also a significant source of greenhouse gases and volatile organic compounds that contribute to climate change and ground-level ozone pollution."[10])

North Dakota was, as always, awash in cognitive dissonance. Because of climate change, average temperatures in North Dakota are 2 degrees Fahrenheit warmer than they were at the start of the twentieth century. Although North Dakota remains bitterly cold in the winter, the number of extremely cold days has been below average since 1980, and average winter temperatures are about 5 degrees warmer than they were 100 years ago. Even the growing season is two weeks longer than it was in Anna's time. The changes meant corn and other warmer-weather crops can be sowed in new latitudes. This had been one of the topics of the journalism fellowship: the fragility of native prairies in the face of not just a widening corn belt, but a warming climate.

There was no getting around it: North Dakota had an outsized dependence on the industry directly responsible for the acceleration of climate change. And yet the state remained a place where, for the most part, people failed to connect human actions to global warming. At the time, more than half of people in North Dakota, an estimated 55 percent, thought climate change was happening, according to the Yale Climate Opinion Maps, which tracks public outlooks on global warming. Yet only an estimated 40 percent of those people attributed global warming to human activity. Another 43 percent attributed global warming to natural changes. Fewer than half, an estimated 49 percent, were worried about climate change, according to the Yale estimates.[11]

Two events best illustrated the cognitive dissonance roiling North

Dakota before my return that summer. In April, a group of Lakota youths began organizing a prayer camp in opposition to construction of the Dakota Access oil pipeline on the Standing Rock Sioux Reservation, on the border of North and South Dakota. Protesters, led by the Lakota young people, sought to forestall construction of the pipeline under the Missouri River. The young people were, along with other activists, fresh off the success of protests against the expansion of the Keystone XL Pipeline, which would have carried oil from the Canadian tar sands to refineries in the United States. Protests against the 1,200-mile Dakota Access Pipeline began drawing hundreds—and eventually thousands—of people to North Dakota to the camps.[12]

It all happened against the backdrop of the 2016 U.S. presidential campaign. Donald Trump won enough delegates to become the Republican presidential nominee on May 26, the same day he was also the keynote speaker at the Williston Basin Petroleum Conference in North Dakota. The timing allowed Trump's speech to showcase his energy policy. (Pro-fracking, pro-Keystone, pro-Dakota Access pipeline and pro-coal.) None other than the billionaire Hamm introduced Trump. Hamm told reporters that America needed "a businessman, not a bureaucrat." Trump, in return, called the Oklahoma oil tycoon "the king of energy." More than seven thousand people crammed inside the Bismarck Events Center to hear Trump, the *Bismarck Tribune* wrote, some wearing red baseball caps with a twist on Trump's signature phrase: "Make the Bakken Great Again."[13]

I was on my way to Minot to report a story about how the town was adapting to climate change after its catastrophic 2011 flood.[14] Specifically, I was curious about what the city was doing with a hard-to-get $74 million federal grant from the Department of Housing and Urban Development. The grant was part of the Obama administration's push toward building what it called "climate ready" communities. The money paid for some flood-control projects, but it was also supposed to help Minot plan for future floods or other disasters with links to climate change. The conditions of the grant required the city of Minot to host discussions about "the impacts of climate change" as well as to "develop pathways to resilience based on sound science."

On its face, Minot seemed an unlikely candidate for such a grant, especially one that required people to acknowledge the dangers of climate change. Minot was—and remains—a politically conservative place where the main industries are agriculture, the oil business, and an Air Force base. Its motto reflected the trifecta: "Provide, Power and Protect."

Lee Staab, the city manager who nabbed the grant, had to tread delicately in a community where people not only had doubts about climate change, but depended on the fossil fuel industry for a living. "I talk quite a bit about climate change, but we always talk about other things," Staab told me. He said he focused on how to keep people safe and how to build more robust and resilient responses to natural disasters. It was harder to talk about a crucial fact: scientists have been warning for years that because the planet is warmer, some regions will experience more extreme rainstorms that can contribute to the severity of events like the 2011 flood in Minot.

Nearly everyone I interviewed danced around the dissonance. Two of the city commissioners I interviewed sold flood insurance and had to rebuild their own homes after the 2011 inundation, but they told me they didn't think humans had anything to do with climate change. Even Adnan Akyuz, North Dakota's state climatologist, told me that climate change was a "subjective topic." He said he thought that a warmer North Dakota meant a future where farmers could take advantage of the longer growing season by growing more valuable crops such as corn. It seemed impossible that the state's own climatologist couldn't acknowledge the role of humans in shaping the weather of the future.

Eventually, I stopped asking the people I interviewed in North Dakota whether they believed in climate change. Instead, I began asking what they were doing about flooding or extreme heat and drought. Even if they didn't acknowledge that the planet was warming or that people and fossil fuels were responsible, they couldn't deny the consequences of climate change or the need to adapt to what was happening right in front of them.[15]

It was an imperfect way of acknowledging reality, I knew. But most people in North Dakota were too tied to the industry to answer the question honestly. Nearly everyone had connections to oil or gas, even if their families didn't receive regular royalty checks. Beyond the people with high-paying jobs in the oil fields, there existed a secondary economy to service the workers, whether they were in retail, restaurants, construction, or a day care center or car dealership. All those people and businesses paid taxes, which helped support towns and municipal services. The vicious, co-dependent circle had no end in an economy dominated by oil and gas.

The entire state benefited from the boom, even in more liberal enclaves like Fargo, where people were more likely to worry about the consequences of climate change. The state built better roads with production tax money it collected from oil companies. The universities improved their campuses with multimillion dollar core sample libraries and other investments. The Mandan, Hidatsa, and Arikara Nation built new tribal government offices and an interpretive and history center, and issued regular disbursements to enrolled members from oil revenue. Towns like Williston built palatial recreation centers with sales taxes from all the new arrivals who were buying goods and services. Federal money helped, too. Williston built a brand-new $273 million airport—subsidized with a federal aviation fund. Minot also spent $84 million expanding its airport, including a new terminal building.

But it was a volatile prosperity for North Dakota, the second-largest oil-producing state in the country. Unlike Texas, which has a diversified economy better able to weather busts, North Dakota was overly dependent on a single industry. Nothing illustrated the turbulence of an extractive economy like the Minot airport. In the three years since I had first arrived at the tiny airport in search of Anna's story, oil prices dropped from $97.18 per barrel to $22.72. In 2013, there were thirteen flights per day in and out of Minot. Three years later, the shiny new terminal was mostly empty when I flew out of it. There were just three flights per day now, all because of the declining price of oil.[16]

I didn't live in North Dakota or depend on oil for a living. At the end of my trip, I could fly home to Washington, DC, leaving the cognitive dissonance half a continent away. Yet I, too, was a product of capitalism. Although I couldn't be certain, I suspected that if the mineral rights were more valuable, if my family were sitting on one of the gushers that Julie LeFever told me about, the money could have swayed me, too. It's possible I would have turned a blind eye to climate change and been even more aggressive about pursuing the windfall my grandfather dreamed of when he signed that first lease in 1951.

17

SANS SOUCI

May 2017

North Dakota crude: *$44.07 per barrel*

I LEANED INTO THE SENSE OF DOOM THAT ACCOMPANIED THE NEW PRESI-
dential administration. It was my job, yes, but I went in active search
of stories of disasters in progress and climate catastrophes to come.
It was as though I was seeking out stories of other people's cognitive
dissonance in order to rationalize my own.

Chris and I still dreamed of being parents, so we kept trying. We
could see no way out other than to push to the very limits of reproduc-
tive medicine and our finances—we even took $30,000 out from our
home equity line of credit to pay for a final stage of fertility treatment. A
confrontation with reality and our expectations was almost inevitable.

In February, I flew to Miami to write about how, in the face of sea-
level rise, speculators were conspiring to buy real estate in Black neighborhoods on higher ground. A month later, I flew to New Orleans
to write about a resiliency conference where emergency planners and
government officials urged people to take the coming consequences
of climate change seriously. No one at the conference bothered to talk

much about the source of climate change or what could be done to slow its inevitable consequences. They were too dejected with the new president's pledge to exit the Paris Climate Accords. Instead, they warned everyone to buy flood insurance.

The Democratic governor of Louisiana, John Bel Edwards, tried to sound a warning in his keynote address: "Just a few years ago, I remember sitting in meetings and they talked about the best-case scenario, the worst-case scenario," he said.[1] "In many cases, what we're looking at today, the worst-case scenario from a few years ago, is the most likely scenario that we face going forward. And this is disconcerting, and what it establishes without a doubt is that we're in a race against time in Louisiana."

I was in my own race against time, manic in my professional pursuit of disaster. In May, the upper Mississippi River flooded, and I traveled to St. Louis. There, I wrote stories about decisions people were making about levees, land use, and flood control, and how it all seemed to be on a collision course with climate change. I thought perhaps if I spent enough time along the Mississippi River, with all its symbolism, I could find one, coherent story to tell about climate change. I figured if I talked to enough people, one river town at a time, I could weave together everything I knew. All these people had the river in common if nothing else. I was born in a hospital in Minneapolis alongside the Mississippi, I had the river in common with them. Mississippi was the only word I knew how to spell before I could read.

In a suburb not far from downtown St. Louis, I interviewed a woman whose house had flooded over and over again.[2] I asked the standard reporter questions: her age, her children's ages. We were both forty-three, and she asked if I had any kids. No, I said, shaking my head, not elaborating. I was too old to respond anymore with a cheerful "not yet!" It was too much information to tell her that, just a week earlier, a doctor had transferred the very last of our frozen embryos into my uterus.

From St. Louis, the highway south is fast and wide as it descends toward lower-lying cotton country. As I drove along Interstate 55 between St. Louis and Cape Girardeau, an extraordinary tingle took over my body. The heightened sensation was so aggravating, so irritating that, as I sped down the highway to arrive on time for an interview in Cape Girardeau, I shimmied out of my bra. The feeling was my progesterone levels spiking, and I knew then that I was pregnant.

I told no one what I knew, not even Chris.

———

A few weeks earlier I had been in New Orleans, at the resiliency conference within view of the Mississippi. I skipped out on the final day of the meeting to drive south toward the southern reaches of the river. "This is peculiar country," Joan Didion wrote in 1970 of Plaquemines Parish, the tendrils of Mississippi River delta that unfurl into the Gulf of Mexico to the southeast of New Orleans.[3]

I read Didion's description at night in the book *South and West,* while I was exploring the river from St. Louis south to Memphis for my story. It was from a slight book of raw notes from an anxious reporting trip Didion made in 1970, but it hadn't been published until just before my own time on the river in 2017.

It's unlikely Didion saw Plaquemines Parish from above, but I did. An environmental group offered me an aerial tour of ongoing marsh restoration, to help me understand the work they were doing to rebuild the coastline in the face of sea-level rise and climate change and decades of human interference in the Mississippi's trajectory and flow. We flew over the place where seventeen square miles of delta give way to the sea each year. From a small plane, much of the pocked marshland looked like rotted fabric, spotted by decay. I saw only vulnerability, and I worried about the cost and futility of the work being undertaken to save southern Louisiana from the river's forces. The cognitive dissonance overwhelmed me. It was clear from above that it was an impossible task.

It's what Didion saw, too, but from the ground. She was there a year after Hurricane Camille, where piles of hurricane debris were as much a part of the landscape as sea and sand. The devastation along the Gulf "had an inevitability about it: the coast was reverting to its natural state," she wrote.[4]

In 1970, Didion wrote that other white women in the South were suspicious of her unstyled hair and utter lack of what might have been considered at the time as "submission" to her husband. She was free to do as she pleased, including going on reporting trips without him. "It occurred to me almost constantly in the South that had I lived there I would have been an eccentric and full of anger, and I wondered what form the anger would have taken," she wrote. "Would I have taken up causes, or would I have simply knifed somebody?"[5]

It was the same question I had about myself in North Dakota. If I lived there, would I have been able to stand up to all the people who wouldn't admit their contributions to climate change, or would I also succumb to cultural pressures? I didn't think I had it in me to knife anyone. But would I have understood that I was free to do as I pleased, to talk out loud about climate change? Like Didion, I, too was an eccentric on this Mississippi River reporting trip. I was an oddity as a forty-three-year-old married woman in Missouri with no children. I could do as I pleased, which included driving along the Mississippi River and asking strangers why they thought their world was changing so much, and what they thought we ought to do about it.

There were other uncanny parallels between Didion's time along the Mississippi and my own. The possibility of growth in small Mississippi delta towns, she wrote, is "ever yearned for, and ever denied."[6] It was a perspective still on display in towns along the river, where commercial and residential development in floodplains served as little more than a cheap boost to the tax base. But they led to massive flooding problems that now required bold intervention or a federal bailout to fix.

I sat talking in the office of one Missouri mayor who, when asked what his flooded St. Louis exurb was best known for, had to think for a moment. It was the sheer number of fast food restaurants in the town, he said. And the bright blue water tower, he added, visible from the highway.[7]

No one would clamor to bail out his unremarkable exurb, I knew. Americans can barely get it together to rescue New York City or Miami or New Orleans from sea-level rise.

I had seen firsthand as a journalist our failures, before, during and after Hurricane Katrina in New Orleans. Government and social institutions of all stripes failed people in front of my eyes. There was no explanation for the failures other than the belief that it is acceptable to leave some people behind. It wasn't just the abandonment of people on rooftops and in shelters, unable to evacuate. It was whole neighborhoods, schools, and healthcare systems and evacuation plans that were subpar by both design and neglect. It was as though no one was in charge of caring.

The stories and the people I encountered in 2005 lingered with me as I visited towns along the Mississippi. In New Orleans for the resiliency conference, the memories of Katrina overwhelmed me, a mental gumbo of the profound and the mundane: *This is where I talked to that man who was turning on spigots outside of buildings to find some drinking water. Here is the gas station where I interviewed the people who didn't have enough money to fill their tanks so they could get out of town. Here is the underpass where I saw a body floating in floodwaters for days, whenever I drove by. Here is where I stood over the body of another dead man, after he'd been shot by police and left to die in the street. Here is where, the morning before the storm hit, I had a last hot breakfast, biscuits spread thick with butter.*

One beautiful thing about New Orleans is that it will allow you to slide for some time, the river making its way to the sea. Then, though, there is a reckoning.

Twelve years after Katrina, seeing from the sky the threat from rising seas, I wanted to grip the people of Louisiana by the lapels. I wanted to confront them like a sheriff in a disaster movie, one of those films where the squad car's headlights illuminate the driving rain that will eventually breach the dam. *Can't you see the peril? Don't you want to outrun it, now, when you have a head start? Go! The water is rising!*

Climate change demands we come to terms with the legacy we leave for those who come after us. No wonder we look away; we cannot bear to see how we have promised over and over again to take care of it tomorrow. We lie to ourselves even as the mildew creeps up the wall and the flared gas burns.

If Didion had seen Plaquemines Parish from above, would she have been able to write the story she left California to do? After reading her notes, Katrina and even Ferguson emerged to me as inevitable. Didion's trip wasn't a failure, she was just reporting a story that hadn't happened yet. Race, intertwined with environmental peril, decades apart.

To my shame, like Didion, I didn't write the story. Or at least, I didn't write the story I went up in the plane over Plaquemines Parish to write. I didn't know how to write about it objectively, without shaking people by the lapels. I couldn't, so I didn't. My editor was not pleased.

The fabric of my own life by then was rotting, its narrative no longer in my control. I had no room in my heart to write stories about other people rushing headfirst into inevitable loss.

In my search for Anna, I took many pregnancy tests in hotels along the highway. All while on assignment as a journalist over the years, all negative. The first time, I peed in a cup at a Hampton Inn in Bismarck in 2014. A day later, just to confirm I wasn't pregnant, I bought another test at a Kmart in Grand Forks, taken at the Hilton Garden Inn. A young couple waited nervously at the end of the aisle for me to pick out a test so they could choose theirs.

And just a few months before my journey along the Mississippi, on the grimmest of Valentine's Days ever, I tested in a cup at a Sheraton Four Points near U.S. 1 in Coral Gables, Florida. It was my fifth failed IVF cycle.

Now, though, I was at a Holiday Inn Express in Cape Girardeau, and the test was positive. In all the times I peed in cups in motel bathrooms along the highway, the time in Cape Girardeau was the only positive. I took a photo of the pink plus sign, and then hid it in a secret folder on my phone so the image wouldn't appear when I showed my editor the photos that would be published with my stories. As always, I wrapped the used pregnancy tests in toilet paper, so the housekeeping staff didn't see evidence of my futility.

I took comfort from Didion's line, written from a Howard Johnson's hotel in Meridian, Mississippi, as she plotted an escape from the South back to California. "Sitting by the pool at six o'clock, I felt the euphoria of Interstate America: I could be in San Bernardino, or Phoenix, or outside Indianapolis."[8] I could be in Bismarck, or Coral Gables, or Cape Girardeau.

In downtown Cape Girardeau, I bought ice cream and wandered along the fifty-four-foot tall metal flood barrier designed to keep the downtown from flooding. Portions of it were painted with images of famous Missourians. I looked for the women, once again secretly in search of a baby girl's name for the list in my phone. There were mavericks and bombshells and writers with uncomplicated Midwestern names: Josephine Baker, Calamity Jane, Ginger Rogers, Betty Grable, Jean Harlow, Kate Chopin, and Laura Ingalls Wilder.

A digital readout at the Merriwether Street pumping station adjacent to the wall showed that the river was at 45.9 feet, its sixth-highest level in recorded history. "If the pumps were to fail, downtown Cape Girardeau would flood quickly," the city noted on its website.[9]

It wasn't until the next day, driving over the bridge to Illinois, that I fully grasped the scale of the threat behind the flood barriers. From above, I could see that I had walked next to a forty-five-foot wall of water in the city's historic downtown. I had even stopped for ice cream as I pondered baby names.

Levees fail. We all know this.

In Cairo, Illinois, I drove twice, windows down, over a perilous two-lane, eighty-eight-year-old bridge that spanned the Mississippi River at its confluence with the Ohio. Twice, just because it was so thrilling to have all that swollen river beneath me, I had to do it again.

In 1803, Lewis and Clark camped at the confluence to learn how to calculate latitude and longitude for their journey. Now, Cairo was a living ghost town, evidence of willful neglect of the Black people who make their lives there, and a harbinger of climate change. In Cairo, I saw confirmation of what I witnessed in New Orleans during Katrina: we have already made choices to abandon some places and some people, even if we don't say it out loud.

In Thebes, there were no flood barriers. The swollen river oozed wherever it found room, creating a broad, fast-moving lake. I yelled at two boys to stop playing in the floodwaters that covered their playground. There was no telling what contaminants were there. It was my first motherly act, and they ignored my scolding. I took a photo of them splashing on a partially submerged seesaw, their brown pit bull frolicking next to them. "Are you from the news?" they asked. Yes.

My first motherly act in Thebes, Illinois.

Toward Memphis, the topography flattened and the river widened. I stopped at a bend in the river known as Sans Souci, where signs said it was once the site of a ten-thousand-acre plantation. It was also the site of "the famous and formidable Plum Point" mentioned in Mark Twain's *Life on the Mississippi*.[10] There, I struck up a conversation with a white man, a retired engineer, who was reading a Nicholas Sparks novel and drinking beer under a picnic shelter. The locals who call it "San Sue" have no sense for the romance or the melody of the original French pronunciation, he told me. "It means without care or worry," he said. I raised an eyebrow, but I kept my mouth shut. Sans Souci was once a cotton plantation, home to fifty-seven enslaved people.

He offered me a beer from his cooler. Who among us doesn't want to read a novel with a cold beer by the river on a pleasant spring afternoon? I lied, reluctant for this man to be the first to hear the news of my pregnancy, even before the father of the child. I don't drink while I'm working, I told him. For a few days along the Mississippi River, I had a secret. I said it aloud only to myself: *I'm pregnant.*

Two and a half weeks later, I exited the Farragut North metro and headed west on K Street toward my doctor's office in Washington, DC. I kept my eyes down, looking at the broad stripes of the crosswalks as I crossed 18th Street, 19th Street, and then 20th. Blood tests showed my HCG levels were doubling daily, as they should be. But my body had no memory of a good outcome after this kind of ultrasound, and I was steeled for bad news. I signed in at the front desk with one of the fake flower–topped pens the clinic used so that people didn't walk away with them.

We were in a narrow exam room. This was our third doctor at our third clinic. Chris once again was in a tight space between the exam table and the wall. Just like the last time, Chris had to lean across my body to peer at the grayscale images on the screen. Our doctor walked in, businesslike but cheerful. She was prepared to give us good news.

Her face turned expressionless as she moved the ultrasound wand inside me. She couldn't find the pregnancy on the screen. Her demeanor grew even more businesslike as she finally identified what she was looking for. She exited the room and brought back an ultrasound technician for confirmation. The ultrasound showed the embryo attached to my right ovary, fluttering with the electrical signal of a heart that would never fully form. It was an ectopic pregnancy, doomed.

No one could figure out how the embryo got there, to such a wrong place outside of my uterus. I don't even have a right fallopian tube.

Chris placed both hands on my forearm, unaware he was pressing down on me. I felt stifled, confined. I wanted to scream "don't touch me!" but I couldn't, not there in that room. Even there, I had to temper my emotional reaction for others. I wiggled gently at Chris to let go of me even though I wanted to fling his arm off. I wanted to leap out of the stirrups. I wanted out. I thought of Anna and her uncontrollable postpartum bleeding in 1907, with no outlet for her rage and fear. I held my breath until I could no longer hold back the spasms. I heaved out a guttural and tearless sob, and then another and another. I sucked in big gusts of air, my diaphragm convulsing uncontrollably, my pale heaving belly exposed at its ugliest right there in front of my husband and my doctor and the ultrasound technician.

My doctor made hurried phone calls to a specialty surgeon at a hospital in suburban Maryland. She forbade me to leave the clinic until my brother-in-law could drive me to the hospital. Just in case my ovary ruptured when I was alone and I bled to death. This time, I was in peril.

18

THE ANTHROPOCENE

September 2019

North Dakota crude: *$62.48 per barrel*

A DECADE AFTER THE FIRST ONE, ANOTHER MANILA ENVELOPE SHOWED up, with yet another lease. Its arrival was just as unexpected, if not more so. This time, though, it was in my mailbox. This time, I was the one depositing the $4,000 check. It was issued by Diamond Resources, a company that manages title searches and leases and other legal paperwork for oil companies. I signed the new lease on my father's behalf, as his conservator. The money went into his checking account, the one I managed for him after a judge in Oregon decided he wasn't capable of looking after his finances.

Like before, I was surprised that an oil company still wanted to drill into the ground beneath Anna's old place. I reached out to Josh Gregoire, the landman handling the lease for Diamond Resources. He emailed me a map of the other wells being drilled in the area and said he expected work to begin on Anna's land sometime during the span of the three-year lease. I still needed to sort out some probate issues if we ever wanted to be paid royalties, just like Derrick Braaten told me years earlier. But eventually,

Gregoire said, an oil company would drill on the land. Eventually, they would pay my dad royalties on the oil they unearthed.

I remained doubtful—until Gregoire told me which company was leasing the mineral rights. It was none other than Continental Resources, owned by Hamm, the billionaire who bought North Dakota.

I didn't want to take Hamm's money, even on my father's behalf. But as my father's conservator, I couldn't turn down the lease or the check that accompanied it. I was legally obligated to act in my father's best financial interests. The lease payment was barely enough to cover one month's room and board at my father's assisted living facility, but he needed every penny. What was $4,000 to Hamm, after all?

Two years before the arrival of the most recent lease, my father's girlfriend texted me in the middle of the night. He was in the hospital in Oregon, she said, after being pulled over by a sheriff's deputy for erratic driving on the Fourth of July. He was driving a brand-new red Mazda Miata convertible, an impulse purchase he couldn't afford. He couldn't really enter or exit the car either, not without the top down. The traffic stop was the fourth time in three days police had pulled him over in the new convertible. All three previous times, the cops gave him a warning to slow down in the new, flashy car. The fourth time, the deputy could tell something was wrong. He called an ambulance to transport my father to the hospital.

I flew to Oregon from Washington, DC, that afternoon. My sister flew in from South Africa a week later. Our dad was hospitalized for the next two weeks, too manic and too delirious to leave. The mania was triggered by a urinary tract infection he developed after prostate surgery—I learned that erratic and delusional behavior is a common symptom of a urinary tract infection in older people with dementia or those who have a bad reaction to anesthesia. A neurologist pulled up an image of my father's brain on a computer. He showed me a healthy brain for comparison. Even I, untrained, understood how my seventy-two-year-old father's brain had retreated, how it had shrunk.

Our father had nowhere to go after his discharge from the hospital. He was too agitated and too violent to be at home. In all the uncertainty, he and his girlfriend of six years split up, and she decided to sell her house and move eighty miles away. We needed to get all of his stuff out, she told us, so she could prepare the home for sale. My sister and I scrambled to find our father somewhere secure to live, even as we dismantled the woodworking shop he had so carefully assembled just a year before.

All of our father's woodworker friends bought the contents of his shop, at bargain prices. The bandsaw, the table saw, the air compressor, the wall of clamps, and a shed full of figured maple. The woodworkers couldn't believe the deals they were getting, couldn't believe the story I was telling them about my dad, or about the red convertible he bought. No one could believe what had happened to the artist they knew, the tall man who liked to tease people. They shied away from him, from the word dementia, afraid it was catching. I shoved wads of cash in the back pocket of my grimy shorts, my heart breaking.

I returned the Miata to the dealership, seething as I negotiated a $1,500 penalty for all the miles my father put on the new car. When we went to the bank to close one of his accounts, his favorite teller fought back tears. "Bill seemed to finally be getting back on his feet after your mother's death," she told us.

There was just so much inconsequential stuff to manage. I drove a five-foot cast iron garden bell to a Home Depot parking lot. A man gave me a crisp hundred dollar bill and I helped load it into his pickup bed. I boxed up kites and sent them to a collector in Seattle, who resold them online. Seven years after our mother's death, we finally found a buyer for her Nancy Ann Storybook dolls. I photographed them in the bright sun, in their Little Bo Peep and Red Riding Hood costumes. "Do their little eyes blink?" a potential collector asked in emails. She agreed to buy them, but she was nervous about the Craigslist handoff. She asked to meet us in the parking lot of the Starbucks at the outlet mall off the interstate. Her husband watched from the cab of his pickup as she handed over $180 in twenties. The dolls are her birthday present, he said.

Steph and I collapsed in giggles as they drove away. The doll collector was afraid of two women in their forties who were selling their mother's childhood toys out of the side door of their father's minivan. We blew the windfall on iced coffee and the J. Crew outlet store.

After my sister flew back home to her husband and her children and her life in South Africa, I loaded the minivan with three thousand pounds of wood and shop materials I couldn't otherwise sell or give away or cram into a storage unit. I drove a final time to the dump, a place with the ominous name of Coffin Butte. For twenty-eight dollars a ton, you could toss pretty much anything you wanted over what felt like a cliff.

Alone, I unloaded all three thousand pounds, winging sheets of plywood and shucking stacks of dusty *Fine Woodworking* magazines into the void below. A woman parked next to me swung a microwave into the pit.

"Feels good, doesn't it?" she said to me.

It did, yes.

Afterward, I drove away from the landfill, gravel dust trailing the van. A few miles away, I pulled over on the side of the road and cracked open the cold beer I had stashed in a cooler for just this purpose. I looked out at the horizon. The air was amber with wildfire smoke. My skin was salty with evaporated sweat. I stank. But I was back in the West, where I needed to be. Where I wanted to be.

By late September, I said goodbye to my life in Washington, DC. We sold our house, and Chris joined me in Oregon. The fertility clinic refunded us the $30,000 we paid in our final, nearly fatal attempt to have a baby. We were done. Disappointment was so familiar it had settled into my features. I didn't like how it felt in the set of my jaw. It went unsaid, but Chris and I both understood: the move was exactly the fresh start the two of us needed. It was time for us to find a way to live lives rich with satisfaction. Had my father not needed to move, the money he made selling his house and his monthly Social Security payments should have been enough to live on for the rest of his life. But dementia and a breakup changed everything. It cost at least $50,000 per year in rent at an assisted living facility, and it would be even more if our father eventually required intensive memory care. As the caretaker now responsible for his finances, I was pretty sure my father would outlive his money.

A small part of me wondered whether oil money might help.

After talking to Diamond Resources, I reached out to Continental to see if they had an update on when they would drill on Anna's land. I still wanted to tag along, still wanted to see what would happen. Unsurprisingly, a spokeswoman for the company declined my request. "Safety is our number one priority and as such, this would not be something in which we would participate," she emailed me. It was a cursory, impersonal no that she probably sent to any journalist who asked. This was the same company, after all, that six years earlier had refused to allow me and other journalists to take photos on our tour of one of their North Dakota facilities, the one that was visible from a housing development.

And so I drove east, across Montana toward North Dakota, my dog, Mojie, in the back seat. I needed to see for myself what was happening on Anna's land.

———

I stopped for a night in Helena, the town in Montana where my mother was born and raised. The next morning, I hiked with Mojie up Mount

Helena, the peak overlooking the city. It was a brilliant early fall day, and we could see Last Chance Gulch, the historic main drag, and the valley where my mother lived as a child.

Four prospectors who failed to find gold elsewhere in Montana stopped in the gulch in 1864 to water their horses. The gulch was the last chance for the prospectors to strike gold. And they did, setting off a frenzy. On news of their discovery, thousands of miners descended on the gulch. This was a story my mother knew well. Old West nostalgia galloped across America in the 1950s, spurred on by dozens of popular westerns on televisions. My mother was raised on tales of Helena's gold rush and Montana's vigilante past, not to mention the *Roy Rogers Show*.

As Helena became a town and eventually the state capital of Montana, the gulch got named Main Street on maps and street signs. It wasn't officially renamed Last Chance Gulch until 1953, when my mother was eight years old. Then, the Chamber of Commerce launched a campaign to brand Last Chance Gulch as a tourism destination. "Every town and city in the U.S. has a Main Street but only Helena has a Last Chance Gulch, a name with the lure of the West," one of the city's boosters argued to the city council, successfully.[1]

Thereafter, Main Street was Last Chance Gulch.[2] Even now, it's possible to take a tour of the gulch on a tourist trolley with room for well-behaved dogs. Some tour operators emphasize the gold rush intrigue, telling gullible tourists that Last Chance Gulch meanders because stray bullets were less likely to find a target. (In reality, the thoroughfare meandered around all the gold claims.) Mostly, though, the tour operators transport tourists through Helena's grand old neighborhoods, telling anecdotes about the copper and timber barons who built mansions to display their wealth. These were stories my mother also knew well.

Mojie and I wandered the gulch, now a pedestrian mall with a ratio of empty storefronts that might have worried the civic boosters of 1950s Helena. I bought a scoop of huckleberry ice cream and sat on a bench, watching Mojie push a cup of vanilla along the pavers.

These stories of chivalrous cowboys, prospectors, and last chances and unexpected windfalls were what shaped my mother. This romanticized version of the West was why she bought lottery tickets, why she played the penny slots, and why she experienced such a sense of exhilarating confirmation when that first check showed up from the oil company in 2009. It was why she believed in windfalls, why everyone she knew did, too, including her daughter. It was why I was sitting on a bench eating ice cream with my dog in Last Chance Gulch, why I went in search of Anna. The stories weren't entirely true, but they continued to blow down that gulch, making believers of all of us even when we knew better.

———

The next day, I drove across the Montana state line toward Williston. I stopped at the North Dakota welcome sign to film oil field traffic barreling around a scenic bend into the grasslands and buttes of western North Dakota. It reminded me of the time Chris and I paused five years earlier to take pictures of a similar sign near the South Dakota border. We had been so hopeful as we collected quirky road names for the future child we were so certain we would have.

As I set up my tripod, a driver in a pickup accelerated past me. The pickup belched a black cloud of diesel exhaust in the direction of my camera—a practice known as "rolling coal." *The men of the oil patch welcome you back*, I thought.

Never mind. It was finally time to make good on my pledge to try out the waterslide I first visited on that subzero day in Williston five years earlier. This time, it was a warm, shadeless day in late September, and I needed to be quick. Not only was I traveling with expensive camera gear, but I had the dog with me. I worried about leaving Mojie in the hot car alone for too long, even with all the windows rolled down in the least exposed parking spot I could find.

At the front desk, I explained why I was there: it had taken five years to fulfill my promise, but I was back. I asked if I could bring a GoPro with me to film it. The woman working the desk grinned, eager to be in on the adventure. She got permission from her supervisor to bend the rules about filming. Go have fun, they told me. (It was, after all, the most important rule of the rec center.)

Two girls about the same age as my nieces greeted me at the top of the slide, curious about my GoPro and how I planned to use it.

"It's her birthday," one of them said, pointing to her friend.

"Happy birthday!" I said. "How old are you?"

"She's ten," the first girl said. "I'm eight."

"Eight is pretty fun," I said. "I did it for a year."

"I don't want to be ten," she said. "It's too hard."

I laughed. This was true. Ten was certainly harder than eight. And nine, well, that was when everything began to get complicated.

The girls slid down, one after another. The lifeguard looked to the pool to see whether it was clear. She gave me a nod, and I shoved myself away, one-handed. I swerved fast, side to side, disoriented by the brightly colored lights flashing inside the tube, but determined to hold the camera upright. It seemed too fast at first, and then it was just right, a thrilling slippery whirl before crashing into a wall of water in the slow-down lane at the bottom. I went down the slide three times more, even though I had all the footage I could ever want. It was just too much fun; I didn't want to leave.

The pool must have been a magical place at the height of winter.

I could imagine the water enveloping all those bodies with warmth and the windows fogging with steam. I could almost hear the lifeguard whistles and the shrieks of all those happy children. It would be easy to follow the rule about having fun.

———

Anna's place.

Once again, I drove toward Anna's. This time, I knew the way. In my absence, the leaning red barn had collapsed—the one with the gambrel-shaped roof that I first saw six years earlier, while looking for Anna's place for the first time.

I drove a little past the turnoff into Larson, to a small cemetery at the edge of the barely there town. Anna's husband, my great-grandfather Andrew, was buried there. It was my first time at the cemetery; I had not known Andrew was buried nearby, not until someone posted the cemetery on Find a Grave, a website crowdsourced by genealogy buffs who take photos of burial sites and enter names of the deceased.[3]

Andrew's small, slate-colored gravestone sat close to the earth, like Anna's in Jamestown. The stone was engraved with his name, his date of birth and death, and the epitaph: "Father." Not husband, just father. I would never know why, but Andrew was too ashamed to bury his wife in a grave beside him.

As I approached Anna's, or at least the stand of cottonwoods I decided six years earlier might have been where she and her husband once had a homestead, I was relieved to see no evidence of drilling anywhere nearby. The land appeared unchanged from my first visit. I could still squint and imagine what it must have looked like the summer Anna lived there in 1906. Insects were abundant, as was birdsong. The wind rustled in cottonwoods still leafy and green from the rainy summer.

The sunflowers lining the gravel roads were flourishing, too, even as they neared the end of their growing season. When young, sunflowers twist and turn their faces to follow the sun in the sky. As the sun rises, they face east. As the sun moves across the sky, the flowers turn west to follow the light toward sunset. All that heliotropic twisting is caused by one side of the plant growing a little taller at a time, fast enough to follow the sun across the sky over the course of a long summer day in the northern Great Plains, but too slow to be perceptible to human observation other than with a time-lapse camera. As the sunflowers mature, they settle into an eastern-facing posture, strong enough from all that twisting to withstand the ever-present prairie winds.

Unlike their cultivated, dinner-plate sized relatives, the wild sunflowers near Anna's homestead have ropy, multibranch stalks. The stalks turn from green to tan as the summer ends. Their flowers turn from golden to husky brown. Even as their palette mutes, the sunflowers stand fixed in their eastern orientation, their sturdy-but-brittle

stalks in dark relief against the brightness of the snow. I had seen sunflowers there in all seasons. Evidence of the passage of time seemed to
be everywhere on this journey.

———

The next day, I drove south to Watford City, a small town in western
North Dakota. I wanted to sit in on a public gathering of the Bakken
Backers, a pro-drilling trade group conducting its annual meeting. The
high school hosting the meeting was brand-new, paid for with oil taxes.
I eased into a cushy seat at the top of the auditorium, an event space
fancier than even some professional theaters I had seen in New York
or Portland.

A decade into the boom, and there at that meeting, industry officials were still talking about the prevalence of flaring. It had even been
the main topic of their board meeting earlier in the day, Ron Ness, the
president of the North Dakota Petroleum Council, told those gathered
in the auditorium. "Nobody wants to flare this resource. We understand the importance of capturing it," he said.

This was a disingenuous comment, at best. If state regulators
didn't want producers to flare gas, they could have denied them drilling permits until they installed methods to capture the natural gas. Or
they could actually penalize producers for flaring. But they didn't. And
the drillers continued to flare, complaining that low natural gas prices
gave them little incentive to build new pipeline capacity or to find ways
to capture the gas. It was more economic, they said, to just burn it as
waste. They had little argument from the state legislature, which a few
years later passed tax incentives worth as much as $6,000 per month
per well to entice drillers to install capturing equipment.[4] They were,
essentially, paying the drillers to act.

Earlier that spring, I read an article by a reporter with the *Associated*

Press detailing the state's continued inability to effectively regulate flaring. North Dakota's legislative research agency told the news agency that in March 2019, drillers burned off 20 percent of the record 2.8 billion cubic feet of natural gas that emerged from wells across the state. It was "enough to heat all North Dakota homes for a month 10 times over," the piece noted, and well above the established flaring limit of 12 percent. All that flaring meant "wasted money and unnecessary carbon dioxide emissions that worsen global warming," the reporter wrote. But then the reporter had an observation that puzzled me. He described flaring in North Dakota as "a picturesque feature of the oil patch, especially at night."[5]

Perhaps we've been conditioned to describe it as picturesque, the stark beauty of man-made objects set off by the endless skies and prairies. But I had photographed plenty of flares, and there was nothing picturesque about the wanton wastefulness of all that CO_2. The flares were a giant belch in the middle of the country. It was climate change happening right in front of us, visible to the naked eye in the form of a hot orange flame.

A flare in North Dakota, I thought, could be the iconic image of the Anthropocene, the new epoch scientists propose adding to the geologic time scale to designate the global impact of human activity.

Scientists for the past decade or so have been working to identify a place where a layer of sediment marks the start of the possible epoch. One of the primary markers for the Anthropocene will likely be the presence of datable nuclear fallout in the layer, Allison Stegner, a researcher at Stanford University, told me. Stenger, a conservation paleobiologist, heads a team of scientists looking at layers of sediment captured in a reservoir built in 1892 in California. She's among eleven teams of scientists examining locations on five continents in search of a so-called golden spike, the official, sedimentary marker of the start of the Anthropocene. For comparison, the previous epoch, the Holocene, began about twelve thousand years ago at the conclusion of the last ice age. Its official reference point is a line of pollen and dust in an ice core sample from Antarctica.

The year scientists are considering the start of the Anthropocene will likely be 1952, give or take a year on either side. That's when atmospherically detectable levels of fallout settled on the earth from nuclear weapons testing. It's the year marking the globalization of war.[7]

It's also the start of the oil boom in North Dakota and the beginning of the wave of Old West nostalgia. The coincidence seemed uncanny, but Stegner told me there were other cultural and anthropological markers associated with the beginning of the global impact of the age of humans. Scientists have deemed these the Great Acceleration, when population, water use, and even markers like international tourist travel, paper consumption and fertilizer use began shooting up on charts.

"A lot of the things we're thinking about with the Anthropocene really connect inherently to culture," Stegner told me. "All of these different pieces seem to really kind of pull together around 1950-ish, right around that time."

The Anthropocene, wrote David Biello in *The Unnatural World*, is an attempt by the scientists who maintain the geologic time scale "to write our species into it." Implicit in the creation of the Anthropocene are two outcomes, Biello wrote: the Anthropocene either progresses with the changes humans wrought, or humanity does not last "and there is no one to see the record in the rocks."[8]

Also implicit in designating the Anthropocene a new epoch is the understanding that because it is a human-created measurement of time, it is also humans who must find a way to address the changes we've wrought to the oceans, the atmosphere, and the earth—or perish from them. "If people, plants, and animals don't like the climate of 2100, 2500, or perhaps even 25,000, they will have us to blame," Biello wrote. "And if human civilization is to persist to see that climate, we need to get busy inventing a better future."[9]

Stegner told me she thought it was important to officially name the human-created era because it underscores the grave threat of our impact on the atmosphere, ocean circulation, and the land surface itself.

"It's sometimes hard to look at what's happening to biodiversity on our planet and not feel despairing, because of the scale of what we're doing and the losses are happening so quickly," Stegner said. "And so for me, I think that naming the Anthropocene and saying 'look, we're in it,' is a way of saying we need to really get serious here, because there's nothing else."

The Watford City high school hosting the Bakken Backers conference boasted of a two-year training program to help feed local students into North Dakota oil field jobs after graduation. I couldn't help but imagine the school curriculum twenty years in the future. Would they teach kids to climb wind turbines to maintain them, for example? Because there was nothing picturesque about flaring on the prairie. It was already grotesque, but the practice would be considered an obscene relic of the past that hastened our future decline. I hoped for the sake of the people who lived in Watford City that they still had a nice high school in twenty years, but also that they had embraced something besides fossil fuels to prop up their economy. I hoped it for all of us living in the Anthropocene.

So much had happened that made me doubt not only my own trajectory, but our collective human progress. As I drove twelve hundred miles home to Oregon, my obligations and unfinished work and anxieties and failures ran through my head: I had lost two pregnancies.

We elected a real estate developer and reality TV star as our president. The planet continued to warm at an unsustainable rate, in part because of methane being flared in North Dakota. My father would outlive his savings.

It all seemed of equal weight, and I tried to summon up my mother's optimism that we would, somehow, string together the money we needed when the time came.

But I also knew that even the small checks from the oil companies to lease Anna's land over the years meant my family benefited from the environmental destruction, the oppression, and the greed of westward expansion and an extractive economy. When we cashed the lease checks, we had a small role in supporting fossil fuel companies that exploited the earth to reap billions, often only for the benefit of a few individuals, and at the expense of the rest of humanity and the health of our planet. I didn't want the money, not *that* money. I didn't want Anna's mineral rights to be my contribution to the Anthropocene.

Someday, my mother told us before she died, *all this will be yours*. It was a promise and a warning. Only I could decide what legacy I wanted to leave.

19

WE COULD BE RICH

April 2020

North Dakota crude: *$11.76 per barrel*

AND THEN IT HAPPENED, THE ULTIMATE BUST.

It began with what the *Wall Street Journal* called a "poorly timed war for market share" between Russia and Saudi Arabia.[1] The two countries released a glut of oil onto the market in the early days of the COVID-19 pandemic. With airlines, cruise ships, and commuters suddenly stuck at home, demand for fuel softened, and by late April of 2020, oil futures bottomed out into negative territory. No one had ever seen anything comparable, not on a global level. "It's like trying to explain something that is unprecedented and seemingly unreal," Louise Dickson, an oil analyst with Rystad Energy, told *Fortune*.[2]

An estimated ten thousand jobs vanished in the North Dakota oil fields. Approximately 6,800 of the state's 16,100 oil wells went idle. Oil production in the state dropped by a third, to about four hundred fifty thousand barrels per day. By May, there were twelve rigs drilling in North Dakota, down from fifty-five in January. Just one fracking crew was operating in the state; before the market crashed, there were twenty-five.[3]

Lynn Helms, the director of the North Dakota Department of Mineral Resources, was uncharacteristically pessimistic. The crash rivaled the state's notorious 1986 bust, he told reporters during a press briefing. "I'm not sure the world is ever going to return to pre-COVID numbers," he said.[4]

Across the country, drillers couldn't pay anyone to take away their oil. In North Dakota, Continental Resources shut in all but a handful of its wells, the oil industry term for turning off the tap. Hamm's fortune sank with the company's stock price. He lost $3 billion of his fortune in March alone. For much of 2020, *Forbes* estimated Hamm was worth $2.4 billion. The seventy-four-year-old tycoon was still very, very rich, but not as rich as his peak in 2018, when his fortune was valued at $17.2 billion.[5]

The *Wall Street Journal* weighed in with a profile: "Harold Hamm, Fracking Pioneer, Faces a Career Reckoning." Two former friends of Hamm's told the *Journal* they weren't on speaking terms with the tycoon anymore, but that they would never count out the oil man behind the Bakken boom. "Harold has never been afraid to do whatever he thought it took," Mickey Thompson, a fellow Oklahoma oil man, told the *Journal*. (Hamm in 2015 sued Thompson for defamation over a Facebook post criticizing Hamm's stance on fracking-driven earthquakes in the Oklahoma oil fields. Hamm eventually dropped the suit, which sought $75,000 for "embarrassment and humiliation."[6])

The *Journal* reported that Hamm cut company expenses as he moved to keep his oil in the ground until he could sell it for a better price. "Mr. Hamm has slammed the brakes on drilling, shut off existing wells and taken other extreme steps," the paper wrote. One of Continental's steps was to beg the state of North Dakota to declare oil a waste product, a rarely used emergency option during gluts. Continental's lobbyist, Blu Hulsey, asked the North Dakota Industrial

Commission to regulate who could pump oil and how much. Hulsey justified the market interference with the argument that oil prices were in a constant state of global manipulation.

"Continental has consistently championed free market principles and causes, but a free market has never existed in the oil and gas industry," Hulsey said at a hearing to weigh the emergency measures. "And as long as countries are manipulating supply based on geopolitical concerns, we will not have free energy markets."[7]

The North Dakota Industrial Commission declined the company's request, a rare lobbying loss for Continental. Nonetheless, the state made it easier for producers to pause production and restart oil wells once market conditions changed. Drillers were allowed to keep wells uncompleted or put them in inactive status for longer than the regulations usually allowed. This had two purposes: to keep producers from pumping up more unsellable oil to bring to the already saturated market, and to keep them from abandoning wells outright.[8]

If it was bad for billionaires, it was going to be catastrophic for the people who worked for them. So, in June, I returned to North Dakota yet again to see the bust for myself. I also thought it was possible that the April crash, coupled with the pandemic, might be a turning point in resource-dependent states like North Dakota. What if, as Helms said, prices and production never returned to pre-pandemic levels? In North Dakota, where the oil and gas industry was forecast to bring in 57 percent of all pre-pandemic state revenue, the consequences could be dire. Or the consequences could be transformative.

I drove to Watford City, a town bustling with optimism when I had visited just nine months earlier to sit in on the Bakken Backers conference at the high school. A downturn could devastate the community, which was responsible for paying back more than $150 million in debt it took on to build new schools, health care facilities, a sewer plant, and other projects as its population grew from 1,744 to 6,207 people between 2010 and 2020.[9]

On my return, people seemed to be gathering in just two places: the daily line for free student lunches at an elementary school and the weekly line for a USDA food bank at the Lutheran church. In the church parking lot, people queued in their cars for boxes of food, diapers, and jugs of milk. Many of the cars had out-of-state plates from other oil boom states. One fifty-four-year-old man from Colorado who lost his job at a pipeline maintenance yard told me he had never before lacked work. He was picking up groceries for his family of four kids, including an eight-month-old baby who needed the free diapers. It was "a shock to the pride, no doubt," he told me. Another forty-nine-year-old woman, whose husband lost his job as a truck driver, held on to a summer gig handing out the free school lunches. Her husband was collecting unemployment, but her family no longer had health insurance during a pandemic. "I wish I could wake up and this was all over," she told me for a story I wrote for *E&E News*.

All of the chaos and uncertainty—and the hunger—seemed to reinforce what I'd heard at the resiliency conference in New Orleans three years prior: we can no longer rely on past trends to predict the future. Those at the conference were talking about flooding and rainfall and extreme heat and drought, but the pandemic emphasized how the wisdom applied to more than weather. It was everything in our lives, all the systems we thought we understood. A worldwide health emergency exposed the already fragile social safety net of American society as being even less robust than most people, especially those with wealth and power, truly understood. How could anyone prepare for disaster or climate change let alone retirement in the face of unprecedented, unpredictable chaos? The pandemic foreshadowed the turmoil we face, even as the effects of climate change intensify in our communities.

From Watford City, I headed north toward Anna's place. I drove to a cluster of oil wells and an old barn atop a ridge a few miles to the south of her homestead. I'd been photographing and filming one of the oil wells there for several years, and I tried to return whenever I was in North Dakota. In late afternoons, sunsets bathed the pumpjack and barn in warm, camera-friendly light. With a long lens, I could capture the pumpjack in front of the faded barn. The juxtaposition represented to me the old and the new of the place—multiple cycles of boom and bust compressed into telephoto layers. The weathered barn felt familiar, as though it were possible that some long lost relation of mine helped build it.

I pulled up in my rental camper van, a COVID-19 safety precaution that allowed me to avoid hotels and restaurants. The van had a ladder leading to the platform that supported its solar panels. I climbed to the top of the van with my camera and tripod to film the now-stilled objects against the movement of the wind in the grasses.

It was usually a noisy oil well. It was so loud that on previous visits, the sound of the pump nearly drowned out the ever-present wind. The pumpjack had a specific, constant mechanical whine that matched the rhythm of the piston moving up and down as it drew oil from the earth. This time, though, the pump was silent, idled by the price crash and the pandemic. The flare was dark, its hiss absent in the soundscape of the windy ridge.

I sat on the roof of the van as my camera rolled, delighted by all the birdsong. The wind blew my hair wild. Farther west along the escarpment was a copse of low trees and shrubs. Something tawny and undulating caught my eye. I'd seen plenty of deer all day, and for a moment I thought the movement was yet another. Deer graze, they bound, and they freeze, but they don't stalk. This was a cat of some sort, probably a cougar. It was stalking a deer. My binoculars were down below. By the time I turned my long camera lens toward the movement, the animal was gone. And yet I was certain of what I saw.

Wherever I drove, there were birds, even more birds than I'd seen or heard on any previous visit to North Dakota. I saw brilliant male pheasants and pairs of mallards and other ducks I didn't know. I heard meadowlarks, the one bird whose song I truly knew. I watched goldfinches in the bushes and other bigger yellow birds—orioles? warblers?—flit from limb to limb. A pair of migratory terns flew low along the horizon, soft gray and white.

When I stopped the van later to make coffee in the driveway of an abandoned house, I watched as barn swallows swooped through the empty windows. It was as though the swallows believed the ruins were for them, theirs to claim. Then, a purple martin or a swift or a swallow of some sort flew into the van. I jumped out and opened all of the doors and windows, shouting at the bird as I shooed it away. It would have been awful to retrieve its tiny carcass out of the van, a terrible, featherweight omen. Later, another purple martin dive-bombed me as I returned to the van. *I saved your cousin*, I thought. *Leave me alone, you ungrateful brat.* The vision I once had of tumbleweeds and deer bedding down in the abandoned recreation center in Williston seemed less an apocalyptic apparition and more a reality.

The state of North Dakota had just approved a program to spend $66 million in federal COVID-19 relief money to plug abandoned oil wells across the state and to reclaim lands contaminated by brine spills and other environmental mishaps. It seemed important to be in North Dakota in person to write about the consequences of abandoned oil wells and a rare opportunity to witness a reclamation program in action.[10]

State officials described the program as a method of keeping experienced oil field workers employed and available until pre-pandemic production conditions returned. This seemed a stretch. The work was temporary and spread out among multiple small operators paid for only a few days' work on each site. It would last just four months, because land reclamation isn't possible in the winter. It would likely keep only six hundred people at work.

But the well-plugging program addressed a pernicious environmental problem, not only in North Dakota. Abandoned wells dot places such as Pennsylvania or California, where extraction began many decades before environmental regulation. In 2020, the Interstate Oil and Gas Compact Commission identified 56,600 abandoned wells in thirty states, but many environmental advocacy groups suspect there are twice as many, if not more. Sometimes called orphan wells, they can contaminate groundwater as well as emit volatile compounds hazardous to human health, wildlife, and the land itself. They also leak methane that contributes to climate change.

When oil companies are done extracting oil and gas from a well, most state, federal, and tribal regulators require them to plug the wells with cement and reclaim the land they disturbed. Even before companies begin drilling, they're required to post a bond; they forfeit the money if they abandon the well. But many environmentalists and even regulators say the bonds are not high enough to cover the cost to address the problems associated with abandoned wells, let alone plug them. It can cost as much as $100,000 to plug a well, depending on its age and complexity. Wells get abandoned for all sorts of reasons, but generally it's because drillers go bankrupt or are careless operators. Sometimes, larger companies will sell off aging wells to smaller operators who can't make them work. And when oil prices drop, the problems worsen. Bonds rarely cover the full expense of determining who owns an abandoned well, let alone who is responsible for remediating it. Orphan wells often sit forsaken, leaking methane and other toxins for decades while regulators sort out the mess.

All those forgotten wells begged the same old question as flaring: If a company couldn't afford to clean up after itself, why did states allow them to drill? Regulators could say no, but they hardly ever did. Instead, there was an elaborate built-in system that practically anticipated someone else—taxpayers—would have cleanup duty. And not everyone in North Dakota was thrilled that federal coronavirus relief money was paying for the cleanup, instead of the drillers. "The responsibility for restoring abandoned well sites should lie with the companies that own them," the *Bismarck Tribune* wrote in an editorial.[11] "This should've been done years ago," Scott Skokos, executive director of the Dakota Resource Council, told me.

I may have been overly optimistic, but I thought North Dakota's well-plugging program could prove an early glimmer of a potential energy transition in America. Environmental remediation jobs were a major thrust of the Green New Deal, the plan by some Democrats in Congress to cut greenhouse gas emissions while addressing economic inequality and environmental injustice in America. And infrastructure bills pending at the time in Congress proposed spending billions for the cleanup of abandoned wells and mines.

North Dakota's Department of Mineral Resources did not want me reporting independently on the abandoned wells. I thought at first it was because of safety concerns. But then they declined to invite me on a state-sponsored media tour of some of the reclamation work. After I badgered state regulators for information about who was doing the plugging, they sent me a 5,000-page PDF document of contractors who bid on the project and suggested I reach out to the people in the files. So I did. I began cold calling, emailing, and texting dozens of companies to find a contractor who would allow me to accompany his crews as they did the work.

Just one of those contractors returned my call: Tom Brooks of

Tiger Well Service in Watford City. Tom called me on a Monday in August and told me he planned to plug a well on a Wednesday. Quick as we could, Chris and I packed my car with a tent and camping supplies. I kissed Chris goodbye and then sped across Washington, Idaho, and Montana toward North Dakota, yet again.

Two hours from the North Dakota state line, Brooks got in touch. State officials were grumbling about me being on site, he told me. My heart sank. I was on a lonely stretch of Montana Highway 200. I had sixteen hours of hard driving behind me. I couldn't write a story about reclamation work without seeing a crew actually plugging a well, I told Brooks. And I had already come so far. I think he felt sorry for me and perhaps mildly irritated at the state of North Dakota telling him how to manage workplace safety. He agreed to let me accompany him.

The next morning, I met Brooks and his wife Kelly at 6 a.m. at their home in Watford City, which doubles as their office headquarters. They outfitted me with me safety glasses and a hard hat. Then, we climbed into Brooks's maroon extended cab pickup, and we started driving toward a forty-two-year-old oil well near the Montana state line.

On the way, Brooks told me how he built his company. Like many, he arrived in North Dakota during the Great Recession, at the beginning of the Bakken boom. He purchased a $7,500 mechanic's truck, loaded up all his tools, and started working the minute he arrived—and never stopped. "I didn't sleep very much," Brooks said. "I never said no." From that initial investment in the truck, Brooks built Tiger Well Service, which he co-owns with his wife and another couple. Most of Tiger's work is in maintaining oil wells with what are known as workover rigs, a scaffold-like repair assembly that sits atop an oil well. "Oil wells are just like cars," Brooks said. "They break down."

When the pandemic hit and oil prices crashed, Tiger laid off about half of its eighty employees. The company idled all seven of its work-over rigs. Tiger stayed in business with a loan from the federal Paycheck Protection Program, even as some of their competitors ceased operations. When North Dakota announced its plans for a reclamation program, Brooks was eager to bid on as many of the state well-plugging contracts as he could. He just wanted to be working again. The crash of 2020, Brooks said, was his eighth industry downturn since he began working in the Wyoming oil fields in the early 1980s. He told me he learned on his very first job that plugging wells could be a valuable skill to keep small oil-field services companies in business during busts. Maintenance and cleanup offered business opportunities others weren't always willing to chase. "Plugging is your savior," he said.

It was dark when we started driving, but as we approached the well, I realized it was not far from the North Dakota welcome sign where the previous fall, a driver in a pickup rolled coal at me while I filmed.

We turned down a gravel road and drove uphill to the well, deep in the Little Missouri National Grassland. It was beautiful country, with the sort of mixed-grass prairies and vistas seen in old Westerns. The grasslands, which surround Theodore Roosevelt National Park, are an overlapping patchwork of one million acres of land owned by the state and federal government and private holders. Interspersed among the buttes and bluffs are hundreds of oil wells, some dating back to the boom of the late 1970s and early 1980s. The Trump administration was about to open two hundred thousand acres of previously untouched land in the grasslands to drilling; a few months later the new Biden administration canceled out that action by halting all federal leases nationwide.

I was dressed in a variation of what I almost always wear for reporting work that takes me outside: green khakis, a long-sleeved chambray shirt to keep the sun off my arms, and lace-up leather ankle boots with flexible soles. It was my uniform. In it I was comfortable and nimble—I had plenty of pockets and could crouch easily to take photos—but I was also relatively anonymous. The outfit drew no unwanted attention, by design.

What did draw scrutiny was my face mask. I would have worn a mask anyway, as we were in the midst of a pandemic. But it was also a specific COVID-19 safety precaution required by the Society of Environmental Journalists, which awarded me a grant to pay for travel and expenses associated with writing a story for *Stateline* about the well-capping efforts. I told anyone who asked: a mask while I was working was no different to my employers than the hard hat, safety goggles, or steel-toed boots required by oil field operators. I had to follow CDC guidelines. But many in North Dakota saw it as a political statement or expression of ideology. When I made eye contact with one of the men at the well site who worked on the cement truck, he spat on the ground next to me and then laughed at my revulsion.

Tiger was plugging a well that was officially known as Federal 3–32X. Because it was in territory where oil and gas leases are overseen by the Bureau of Land Management, both state and federal inspectors had to sign off on the plans to plug it. State records showed the well was drilled in 1978; in its lifetime, it produced 221,587 barrels of oil. It was a conventional well that descended 9,572 feet vertically into the earth. Most of the wells the state of North Dakota was paying contractors to plug in 2020 were older, conventional wells, and not newer, horizontal wells drilled in the fracking era.

The state assigned Brooks the well at random, but he told me he was familiar with this one. He had worked on it in 2012 for its most recent operator, Alturas Energy of Idaho. Back then, Brooks was hired to figure out why the well wasn't producing. It was too costly to make repairs to return the well to production—and Alturas never paid Brooks. Before Alturas went out of business, the company encountered other problems in North Dakota, according to news reports. In 2015, the state fined the company $900,000 for spills in western North Dakota. Brooks bid $14,650 per day to plug Federal 3–32X. He estimated it would take a week and cost the state $102,550 to remediate.

While figuring out how to seal Federal 3–32X forever, Brooks determined how he could have fixed it. I suspect it bothered him that the well was being plugged, instead of repaired and returned to service. He saw the well as a valuable resource that could make someone money if it were used, and not one to be left behind in the ground. But he was getting paid to plug it, and that was all that mattered, especially when his rigs and his employees had been idle for so long.

The workover rig was already in place when we arrived. The white tower shot into the sky, stabilized with guy wires tied with safety flags so workers didn't run into them. Plugging a well is a relatively straightforward process that, like house painting, requires extensive preparation and clean up. Once the workover rig was in place, Brooks had to diagnose the problems with the well and come up with a plan to plug it, and then begin pumping cement. The cement doesn't set when temperatures drop below a certain threshold, so in North Dakota, the window of opportunity to plug a well is limited to the warmer months.

The job site was noisy. The diesel engine that powered the rig chugged with effort as the crew inserted tubing into the well. When the engine was engaged, the flap on the rig's red exhaust pipe chattered with black fumes. It reminded me of the engine in the children's book, *Mike Mulligan and His Steam Shovel.* Tubing clanged as the crews inserted the pipe lengths into the well, sometimes banging at the connections with a sledgehammer to ensure they were in place. Eventually, a cement truck pulled near, too; its engine noise and the continual roar of the pumps added another layer of sound. After the tubing was no longer needed, the crew pulled it out of the well one length at a time. One of the workers physically walked each length of pipe to a storage rack, releasing it with a loud clatter from his shoulder. The pipes dropped noisily with a predictable rhythm that wormed into my brain: clang, clatter, roll, settle.

Midday, I rode with Brooks and his wife to buy snacks and use the restroom at a gas station in a section of the grasslands with better cell phone service. As he drove, Brooks was constantly on the phone, solving mechanical and logistical problems for his crews. "Make it quick," he snapped at one caller. "I've got nothing but mayhem and confusion right now." He was a problem solver with a mechanic's mind. Even when he complained, it was clear he liked moving around the puzzle pieces of his employees, subcontractors, and jobs. It occurred to me that, when he called while I was en route to North Dakota, it was during one of his mobile administrative windows. My presence was just another logistical challenge to sort out that week. For someone who rarely said no to work, it was easiest for Brooks to tell me yes.

On the way back to Watford City, Brooks stopped at another site where Tiger crews were at work cleaning up a disposal well for oil field waste. Kelly Brooks, who did all the hiring for Tiger, told the workers they were free to talk to me about what it was like working in the oil fields during a downturn. One of the truck drivers sought me out. He was a muscular man who looked like he was in his early forties. His questions came fast, partly because he was an intense guy, but also because he was emptying a storage tank of waste materials into his tanker truck, and he had to keep an eye on it so it didn't overflow.

Are you a writer? he asked. What kind of writing do you do? Do you do editing? How much would you charge for editing my book?

"What's your book about?" I asked.

"Toxic masculinity," he said.

"Ah," I said.

It was rare to run into a man on the oil fields of North Dakota who, within minutes of meeting you, begins speaking with any sort of self-awareness about toxic masculinity—the embedded American cultural message that to be manly, men must be aggressive and cannot express emotions. It was far more likely as a woman to *experience* acts of toxic masculinity, like the guy who spat on the ground in front of me or the driver who rolled coal while I filmed the year before. Or the executive in June in Watford City who hated that I arrived for an interview at his bank wearing a mask, his annoyance shifting to hostility when I declined to shake his hand three months into the pandemic.

The next bit from the truck driver came out in a rush. His name was Nick Chontos. He was once addicted to meth, and seventeen years ago, got sent to prison. There, he turned his life around. Now, he drove waste disposal trucks in North Dakota. I found his urgency compelling. Somehow, Chontos recognized me as someone who would not only know about toxic masculinity but as a person who would understand how he couldn't envision a future that didn't involve telling his story. I gave him my card and told him to keep in touch. On the back of my card, I wrote down "North Dakota Humanities Council," and suggested he look up their memoir writing and self-publishing workshops. I also thanked him for his vulnerability, because it's intimidating to approach a stranger—and maybe the only journalist you've ever met—to ask for help with a creative project. The class and gender dynamics alone were substantial barriers. I suppose it meant he really and truly had eclipsed the toxic masculinity he said clouded his early life. Or maybe like all people who thrive in the uncertainty of boom and bust economies, Chontos learned to jump at opportunity when it presented itself.

The Tiger Well Service crew at work near Watford City, North Dakota.

A few days later, it was time to officially cap Federal 3–32X. I met the welder, Nathaniel McGuire, at his shop near Watford City. I followed his truck, with its Wyoming plates and the chrome sexy lady silhouettes on his mud flaps. When we arrived, the workover rig was gone, already en route to the next job Brooks put together while driving down the highway. It was just McGuire and me and his partner on the job, Freddy Magana. It was much quieter work, until a separate trucking crew arrived to clean out the storage tanks at the site. The foreman asked me to wear ear plugs; it was noisier even than the day the cement truck and workover rig were operating. With the earplugs in, as well as the safety glasses, a hardhat, and my face mask, my senses felt diminished. I worried I wouldn't capture the action accurately, so I set up a time lapse video on my phone to watch later.

McGuire dug up the earth around the wellhead with a backhoe. Then, firing up a blowtorch, he worked in a circle around the wellhead to cut it off at the base. Magana attached the wellhead to the backhoe shovel with a chain. He lifted it up out of the shallow pit and hauled it away, swinging it a few feet above ground. It looked no different than a sewer pipe. McGuire worked fast. He had already made a steel monument plate to cap the well. I took a photo of him holding it, his welding helmet pushed back on his head. Then he jumped back into the pit and, sparks flying, within minutes welded the monument atop the capped well. The monument, a square plate about eighteen inches on each side, was marked with the name of the well and its file number and date of its closure, just in case anyone ever dug into the earth and found it. McGuire snapped pictures of his work with his phone and texted them to Brooks. Then, he covered the well with dirt. It was just like a grave. Eons from now, future geologists might find the monument, its presence in a layer of earth yet more evidence in the geologic record of the Anthropocene.

Eventually, a separate crew would finalize the reclamation process by hauling away the gravel for the road leading to the well. They would restore the land on the surface with a special seed mix of native prairie grasses made for North Dakota. The state would sell off or scrap salvageable equipment, including pumpjacks and storage tanks.

That afternoon, Nick Chontos, the truck driver I'd met a few days earlier, emailed me to tell me he had already spoken with someone at the Humanities Council about helping him with his writing project. He was just one guy driving a waste disposal truck in North Dakota, getting in touch with his feelings. Federal 3–32X was just one well plugged, in a nation of tens of thousands of abandoned wells and mines and toxic industrial sites. But both events felt encouraging. Out there in the Badlands, where both the rock formations and the oil infrastructure stood in stark relief against the big, open skies, for a moment it felt like I was documenting the beginning of the end of fossil fuel use in the Anthropocene.

A few months after my story published, events dashed some of my optimism. North Dakota couldn't spend all of its federal pandemic relief money set aside for well-plugging and reclamation, not by the end of the year. So, in October, lawmakers authorized a program that would give $16 million of the remaining money as grants to oil companies. The money could go toward fracking new wells that companies had already drilled but had chosen not to complete during the downturn while oil prices didn't make it worthwhile. Nine companies took advantage of the program, *Inside Climate News* reported. State records showed that Continental Resources received $5.4 million.

The grants were a baffling corporate giveaway, said Skokos, of the Dakota Resource Council. Why give money to oil companies that were going to frack the wells anyway? "They're giving taxpayer dollars to the oil industry to frack wells with the hope it will bring the state more taxpayer dollars," he told *Inside Climate News*, "rather than taking the taxpayer dollars and actually using it to benefit taxpayers."[12]

I still wanted to imagine the future I envisioned that day on the butte, when I watched McGuire weld shut an oil well. There was such a future possible, I was convinced. I could imagine men like McGuire driving the country, welding shut old oil and gas wells in North Dakota, Montana, Wyoming, and Colorado as well as New Mexico, Oklahoma, Texas, California, and Pennsylvania. Turning the welders and truck drivers and roustabouts of America toward reclamation seemed a fitting evolution of the stories we've been told about the building of this country. It was good-paying, vital work, even if, like always, working-class men were the ones left to clean up messes made by even better-paid guys, the ones who never had to get their hands dirty and who never had to worry about sleeping in a Walmart parking lot.

20

WINDFALL

June 2021

North Dakota crude: *$60.53 per barrel*

SOMETHING LURED ME BACK, PERHAPS THE SAME WINDS THAT WHISPERED to me in the years after my mother's death. They were telling me I was not quite done. And so, I drove to North Dakota one final time in early summer.

I arrived from the west, skirting Williston to head directly toward the place in Burke County where Anna lived so briefly, so long ago. As I neared the escarpment that rises south of the land she homesteaded, an unexpected sight appeared on the horizon. It was an array of wind turbines. Dozens. Hundreds? So many, stretching for so far, I couldn't count them all.

I had photographed this place so many times, beginning with one of my first days in North Dakota eight years prior. I knew the area well. The turbines had not been on the ridge when I last drove through, exactly one year earlier.

An unidentifiable, euphoric emotion vibrated through my body. Overwhelmed by sensation, I pulled over to the side of the gravel road, right where I always stopped for photographs of the barn and the pumpjack. I exited the van. Intuitively, I crouched down and put my hand on the earth. I paused there for a moment, taking deep breaths to ground myself. The wind farm was not an apparition. There, with my hand on the ground, I put a name to the emotion I felt. It was hope. The faded barn and the mustard-colored pumpjack, which stacked up so nicely with a long lens, would now have a third layer behind them in my photographs and videos: a wind turbine.

The wind farm in Burke County, North Dakota.

The turbine in my photo was 485 feet tall, including the full length of one of its blades. (For comparison, the Washington Monument is 555 feet tall.) There were 74 turbines total in the wind farm, generating 193 megawatts of power—enough to power an estimated 34,000 homes per month, according to the U.S. Geological Survey. The turbine was part of a $300 million wind farm that sprang up on the ridge in 2020, after the North Dakota Public Service Commission granted it approval.[1]

Northern Divide Wind, a subsidiary of the Florida power company NextEra Energy Resources, owned the wind farm. A spokesman for NextEra told me the company owned sixteen wind farms in North Dakota, and more than one hundred twenty across the United States and Canada. The company sold the output from the wind farms to electric utilities, municipalities, and power cooperatives. With consistent and strong winds, ample rural land, and good access to electrical transmission lines, North Dakota had all the successful elements for wind farms, NextEra's spokesman said.

That said, it took several years for Northern Divide Wind to get permission to build the wind farm in Burke County. Their first application was denied by the North Dakota Public Service Commission for failing to take into account the unique bird life of the Northern Missouri Coteau region. "The unique grassland and wetland mosaic of the Missouri Coteau not only provides essential resources for many resident species, but functions as important stopover habitat to numerous wetland-dependent migratory birds," the North Dakota Game and Fish Department wrote.

Nothing, it seemed, was simple about a future free of fossil fuels. Eventually, though, the Public Service Commission approved the wind farm. Northern Divide reduced the area of the project by 52 percent, and relocated forty-four turbines to places with less impact to wetlands and wildlife, including what wildlife managers described as a crucial Northern Pintail Duck breeding habitat. None of the turbines were located on unbroken native grasslands.

I drove down the hill from the wind farm, toward the land Anna once homesteaded. Booms and busts shaped this land, but it also had a long history of sustenance for the people hardy enough to make it home. It began with the Indigenous people who first inhabited the land and hunted its bison, and then the homesteaders who farmed wheat and flax and carved out the coal and gravel. Then came the oil leases in the 1950s, like the one that paid for my mother to go to college, followed by the boom that brought me and so many others to North Dakota. Now, there was a wind harvest. Wind easements pay $50,000 per year or more to property owners, depending on how much energy the turbine generates. Local governments also reap substantial property tax revenue from wind infrastructure.

Down at Anna's, I could see the turbines on the ridge to the south. From afar, the rotors looked spindly, their mass diminished by the expansiveness of the landscape. As the day drew to a close, the turbines blinked with red aircraft warning lights. I could no longer squint and see how the land looked as when Anna viewed it. But with my eyes wide open, I could see the future.

This time, there would be no new manila envelopes. The lease I had signed on my father's behalf was about to expire. I checked in once again with Josh Gregoire, a landman with Diamond Resources, the company managing the lease for Continental Resources. There were no plans to renew the lease, Gregoire told me. Continental had shifted its focus, once again, to prospects in the south, in more promising parts of North Dakota. "If they do head back up in that direction we would definitely contact you immediately," Gregoire told me in an email.

It was just as Julie LeFever said when she pulled the core samples for me five years before at the lab at the University of North Dakota: "Not much success up in this area at all." This time, I didn't care. I didn't want any oil money.

From Anna's, I aimed the van diagonally to the southeast toward Jamestown, traveling the same route as the train that took her to the asylum in 1907. I drove to the cemetery, exited the van, and walked toward her grave. As with the first time I visited, I heard the distant hum of traffic on the highway. There was the buzz of insects and the warble of a western meadowlark. Above, a hawk circled, soaring on the air currents. As always, I brushed grass clippings and dead leaves from Anna's gravestone. Then, I placed a bouquet of yellow roses atop it.

There at Anna's grave, I came to a decision. I decided to give up any claim to the mineral rights.

I had been thinking about it for several years, ever since I signed the final lease with Continental Resources on my father's behalf. As my father's conservator, I had to turn in an annual accounting to the court of how I spent his money. My father didn't have much money to spend, so it was a straightforward, albeit time consuming, annual exercise in paperwork. I was allowed to give charitable contributions that aligned with my father's values—I learned this from the training video I was required to watch to serve as my father's conservator. Acting on his behalf, I could sign over the deed to the mineral rights.

I consulted with my sister, who would split the rights with me one day, after our father died. Steph agreed, easily. She had no memories of North Dakota, and she told me she had no emotional attachment to the mineral rights. Neither of us wanted her children to be tied to them one day, either.

I reached out to Derrick Braaten, the attorney in Bismarck who looked over one of my father's past leases. Did anyone ever say no to mineral rights? And could they?[2]

Not really, Braaten said. And even if we said no or did nothing with the mineral rights, an oil company could still drill there because of forced pooling. That's the practice that organizes the oil patch into 1,280-acre grids known as spacing units, and that also limits how many wells can be drilled to draw from each unit. Oil companies must give notice and offer leases, but they are allowed by law to drill even without the consent of all the owners of the mineral rights in a spacing unit—or the owners of the surface rights.

"You really don't have any ability to say no as a mineral owner, unless you own all of the minerals or the majority of the minerals in a spacing unit," Braaten said.

Braaten also told me about someone he knew who inherited mineral rights in North Dakota and gave them to an environmental organization. Like my sister and I, the donor didn't want money from an oil company. Like us and nearly all mineral owners, the donor had no say in whether the land would eventually be drilled. The board of directors of the organization wasn't sure it should accept the money at first, but after consideration, it decided to devote the proceeds from leases and royalties to climate change education.

There were other examples. I read about Terry Tempest Williams, the writer and activist who purchased leasing rights to 1,120 acres of federal public lands near her home in Utah.[3] She and her husband bought the rights as an act of protest against the leasing of the lands to oil and gas companies planning to drill for fossil fuels, she wrote in the *New York Times* in 2016.

"Our purchase was more or less spontaneous," Williams wrote, "done with a coyote's grin, to shine a light on the auctioning away of America's public lands to extract the very fossil fuels that are warming our planet and pushing us toward climate disaster."

Eventually, the Bureau of Land Management revoked her lease, saying leaseholders needed to use leases or lose them. Even so, I appreciated the cheeky way Williams carried out the protest within the structures of the federal leasing program. It was an effective tactic that illustrated the rot at its heart. It showed how it was a system built for extraction, at a time when we knew next to nothing about climate change. It no longer made sense.

The system of severable subsurface rights also seemed to me a failure of property law, an archaic construction as obsolete as old housing covenants that restricted ownership by race. The history of the American West represents "not a simple process of territorial expansion, but an array of efforts to wrap the concept of property around unwieldy objects," the historian Patricia Nelson Limerick wrote in *The Legacy of Conquest,* written in 1987, but prescient for its focus on the strain of fracking on Western communities.[4] Unwieldy as the system was, it was the one we had until we demonstrated its inequity and we replaced it with something better.

So if we could trace the mineral rights all the way back to Anna, the homesteader who claimed the land in 1906, why stop there?

I made a plan: I wanted to confer the mineral rights to the descendants of the Indigenous people who inhabited the land before Anna and Andrew. If the mineral rights belonged to anyone—and I had doubts about individual ownership of the earth beneath us—they belonged to the descendants of the people who lived on the land first, before settlers like my great-grandparents arrived.

There were many ifs in all this, including the complexity of tribal law concerning mineral rights. But if I were able to pull it all off, it meant Continental Resource would have to negotiate any future lease with tribal governments. That was *if* the company ever came back to Burke County to drill where Anna once lived. If they did, it seemed a fitting reparation for a company owned by a billionaire who built his fortune in Oklahoma and North Dakota, places once designated on the maps of North America as Indian Territory.

It would be difficult but not impossible to make it work. I would need first to ask whether the tribal governments in question, the Fort Peck Assiniboine and Sioux Tribes, even wanted the subsurface mineral rights or could accept them. Some might not. But the tribes could decide what meaning and value subsurface rights had within the context of their cultures and histories. If they wanted them, the tribes could use the mineral rights in whatever way they saw fit.

There were no active leases on the land and no plans to drill. It meant I was giving away something of symbolic value, not monetary value. The mineral rights also remained in a legal tangle. Before I could confer anything, I would also have to finally do what Braaten had advised years before: have a court in Montana probate my grandparents' wills. Then, the court would transfer the mineral rights to their heirs: my father, my aunt, and the ten children of my two half-aunts. After all that, I could use a conservator's mineral deed to transfer my father's share.

I knew it would be pricey. If I paid a lawyer, it could cost $10,000 to $15,000. I didn't have that sort of cash to spend on symbolic paperwork and I likely never would. But I knew my way around a county courthouse. I thought it was possible that with a little training, I could learn how to file the necessary documentation myself. Maybe I could also persuade a cousin I barely knew and ten longlost half-cousins to join me. Perhaps other descendants of homesteaders would be intrigued by the idea of letting go of something that, at least for me, had come to seem an inheritance that was never mine to claim.

My windfall was the freedom to set out in search of Anna's story—not what had to be extracted at great cost from beneath the earth.

There at Anna's grave, I thanked her out loud. It was Anna, after all, who made it possible. Her mark on the world was slight, and yet it was she who did all of this. It was Anna who whispered her legacy into the wind.

We could be rich, I thought, smiling, as I walked away from that windy, mournful place. *We could be rich.*

ACKNOWLEDGMENTS

A BOOK EIGHT YEARS IN THE MAKING OWES SO MUCH TO SO MANY PEOPLE, but I'm especially grateful to my writing group, the Left Coast Authors. Nora Brooks and Arianne Cohen made me feel at home as a writer in Portland, and along with Gigi Rosenberg and Danielle LaSusa, always made time for thoughtful feedback on early drafts of so many chapters.

I'm also grateful for the unflagging support of my agent, Jessica Papin, a champion of stories from the West. My editor, Anna Michels, knew exactly where just one sentence would do. Thank you to the entire team at Sourcebooks.

Thank you to the Institute for Journalism and Natural Resources, which got me back out on the North Dakota prairie when I needed most to return. A special thanks to Frances Backhouse and Tina Casagrand, who listened to me ramble on a long bus ride through North Dakota, and who helped shaped this story in a way that only the best writing teachers and editors can.

I couldn't have completed the first draft of *Windfall* without a residency at Playa or my chats with Michelle Goodman at Summer Lake Hot Springs. And Miriam Gershow, thank you for the Big Revision Energy of Cabin 10. I met my deadline thanks to a generous, well-timed residency at another special Oregon place: Pine Meadow Ranch Center for Arts & Agriculture in Sisters. I owe a big thanks to fellow resident Roger Peet for the kitchen conversations that shaped the final chapter of the book.

I'm grateful to the generosity of historians Josh Garrett-Davis and Patricia Nelson Limerick, who helped me understand how the western as a genre shaped the narrative of the Bakken oil boom.

Thank you to Eline Gammelsæter for help with Norwegian translation, as well to fact-checkers Amy Dalrymple, Ramona DeNies, Amanda Kay Rhoades, and Lindsey Van Ness for casting careful eyes on several chapters. Jacqueline Keeler, a Diné/Ihanktonwan Dakota writer, gently pushed me to be more precise. Any errors or omissions are mine.

Thank you as well to Justin Hocking, who helped me look for windows, and to Brent Walth, for a careful read of early versions of Chapter 3. I'm grateful for Jane Gerster's reading of later iterations of the chapter. The joke in Chapter 9 is thanks to Susan Elizabeth Shepard. And thank you to Wes Pope, Julie Perini, and Reed Harkness, who all helped me think more cinematically.

To Jeannine Ouellette and the writers of Elephant Rock 5: Thank you always for our healing time together on the Island of Happy Days.

There are dozens of people to thank in North Dakota, but I'd like to offer a special note to Becky Jones Mahlum for her friendship and her invaluable connections and kindness. Amy Sisk of the *Bismarck Tribune* texted me valuable background on the wind farm in Burke County, even as I stood there gawking at it.

I couldn't have traveled safely to and from North Dakota during the pandemic without a grant from the Society of Environmental Journalists. And thank you to *E&E News*, which funded a good deal of the early research for this project with freelance opportunities—and to Lisa Friedman, for your friendship and professional support.

Thank you to Lesley Clark for your friendship, and for suggesting at the start of this project that I pay attention to *how* I did it. To Kelly DiNardo and Amie Parnes, thank you for critical connections, advice, and encouragement early on and then again when I needed it most. Bridget Matzie offered early guidance on the ideas behind *Windfall*, and Trish Daly helped me identify the title. Thank you, Shauna Dillavou, for listening while I figured out the structure of the book was tied to the price of oil. A big thanks to Stacey Cofield, the best cheerleader and friend a writer could ask for since 1992. And thank you, Noelle Straub, for being such a good friend and listener. I miss living around the corner from you.

I wrote *Windfall* with one of my favorite readers in mind: my sister, Stephanie Bolstad. Thank you for allowing me to tell this story about our shared inheritance.

One of the pleasures of writing this book was reading out loud what I wrote to my husband, Chris Waldmann. *Windfall* wouldn't have been possible without his love, patience, and culinary skills. Thank you, Chris, for finding the weirdest roadside places and restaurants wherever we go. But mostly, thank you for always believing in me.

READING GROUP GUIDE

1. Erika's family has a theory that windfalls show up when they are most welcome (such as the first mineral rights payment right before Christmas in 2009). What theories does your family hold dear? Do you have an example of one in action?

2. Because of split land rights, many drilling decisions are made by individual people. How does this interact with the American ideal of independence? Should anyone be able to make a unilateral decision on something that has consequences for larger communities?

3. How would you describe Erika's relationship with her mother? How does the search for Anna's story bring Erika closer to her mom even though it began after her death?

4. What did you think of including Anna's story with the broader

narrative about Erika's inheritance? How does the history of Anna's homestead claim and eventual confinement shape your understanding of the rest of the book?

5.   When Erika meets with scientists at the US Geological Survey, she learns that the odds of her land producing a profitable amount of oil are negligible at best. At what point does optimism become misguided?

6.   Though the 2008 update doubled the estimate of recoverable oil in North Dakota, oil companies expected (and assured their stakeholders) that the formation was still severely underestimated. What motivates them to speculate so wildly?

7.   How does the inevitable bust of the North Dakota oil boom actually accelerate extraction? What effects does that have on the surrounding areas?

8.   The signage at the Homestead National Monument in Nebraska struggles to reconcile both admiration and criticism for the scrappy homesteaders that proved their own claims while displacing Native American populations. How does that equivocation compare to the version of history presented at the Medora musical?

9.   Despite its obvious effects around them, Erika discovers that many North Dakotans are unwilling to directly discuss climate change. Can we understand anything deeply while only discussing it obliquely? What do we miss when we discuss extreme weather and natural disasters without acknowledging their roots in climate change?

10. At the end of the book, Erika strongly considers granting her mineral rights to the Native American tribal nations that preceded Anna's claim on her homestead. What is the value of this symbolic gesture? Have you heard of any other interesting "Land Back" movements?

A CONVERSATION
WITH THE AUTHOR

Throughout the book, you struggle to reconcile your knowledge of fracking and its negative consequences with your desire for the rights on Anna's land to pay out. Were you surprised by how tempting the mineral royalties were?

Given my mother's belief in windfalls, I wasn't surprised I was tempted—all of us in America are products of capitalism, after all! We've all heard the whispers of riches calling to us, right? But I was a little disappointed in myself at how easy it was to be swayed by greed, especially given what I know about climate change.

From my very first research trip to North Dakota, I began to see how easy it is for individuals to rationalize the effects of drilling when they're getting royalty checks from oil companies. For communities, the rationalization is about their tax base and jobs and paying for fancy recreation centers with water slides. And yes, we all need to put food on the table and a roof over our heads. But at what cost? It really was shocking to see all those people lined up at the food pantry in Watford

City in 2020, in a state that made billionaires of a handful of men whose main skillset was knowing where to drill holes in the ground. I wanted with this book to examine how the promises of riches—often from extractive industries applying immense financial pressure and their advertising might—shaped me and the wider American story of conquest and extraction and greed.

I came to understand that I would rather be a part of changing the narrative around what it means to be rich. Perhaps in the future we will no longer measure riches by the price of a barrel of oil, but instead it could be an index that maps the biodiversity of our ecosystems or the quality of our air and water.

Why was it important to you to address Anna's confinement in addition to the history of her claim on the homestead? What do you hope readers take away from Anna's experiences before and during her confinement?

Anna's story was all too common for women of the time, and it is, of course, even sadder because her mother also had been written off as useless to society. Symbolically, Anna's story in the book represents how willing we are to cast off certain kinds of people, places and even environments in the face of what is widely considered "progress" in America. I hope the takeaway for readers is that we don't need to hide from the shameful parts of our shared, national past. Instead, we must acknowledge them truthfully, and make amends for them by building a better version of this country.

You write candidly about everything from Anna's confinement to your own struggles with fertility. Do you ever feel uncomfortable exposing your life on the page? How do you handle that feeling?

I never felt as though I could leave my infertility story out; it felt intrinsically tied to my search for Anna. When I set out to find out more about Anna, I was grieving the loss of my own mother and how we wouldn't have an adult relationship. I was also trying to have a baby. At the same time, I was uncovering more about a great-grandmother who was forced to give up her role as a mother—and who barely had a mother of her own.

I did cut a chapter with some of the grimmest details of my own infertility, which had an account of the worst moment Chris and I had in our marriage. We saved that for a very skilled therapist. Not every personal detail is in this book—learning to maintain boundaries is critical for any memoirist's mental health.

But there are times I think readers want to know what it felt like, physically, to be in a place or a situation, and I erred on the side of exposure. This is sometimes a tricky line to walk, but memoir should go beyond journal excerpts or a gossip session with friends who enjoy oversharing. Memoir is a constructed and curated accounting of what happened, told in the service of reckoning with or making meaning of larger truths. In this book, my personal story of what happened while searching for Anna helped me as a journalist tell the story of the larger truths about the consequences of American greed and the myths of the West. These truths form our collective American inheritance. The best, most relatable way I could think of to tell the story of the larger truths was by sharing how they intertwined with my own story.

The most challenging part of sharing my personal story was setting aside the voice-from-nowhere that served me so well as a journalist for many years. I have always told other people's stories—and still do. But I had to teach myself to dig deeper, as well as to treat my story with the same care I would give to the people who I ask as a reporter to share their lives with me.

Your research took many years. What did your process look like when you finally sat down to build it into a single narrative?

About two years into the research, when oil prices were crashing, I had an a-ha moment while talking to a friend: The narrative needed to be tied to the price of oil. Readers who like structure might also notice I leaned heavily on classic eight-sequence, three-act screenwriting structure to organize the book.

And even though I had much of the book mapped out before I wrote the full version, many chapters changed in the process of writing them. I wrote a complete first draft in 2019, but there are pieces of some chapters that date to 2015. The process of writing and revision revealed understanding that I wasn't capable of articulating in the planning stage. (Writing is magic!) The book went through many revisions in 2021, deepening the layers and themes. In fact, I had planned to end the story with an account of a visit to Anna's grave in 2019, before something drew me back to North Dakota and I saw the wind farm—and revisited her grave yet again in 2021.

Note also that Anna is in each chapter of the book, even if it's just a sentence or two. This was by design. I had a summary of all 20 chapters up on my office wall, on a scroll of white poster paper that helped me visually understand the linear nature of the narrative. Anna was represented by a teal sticky note in each chapter, as a reminder that her story was the thread necessary to tie it all together.

What are you reading these days? Are there any particular authors who inspire your own work?

Prairie Fires by Caroline Fraser was a major inspiration for *Windfall*, for how it re-examined the myths at the heart of the Laura Ingalls Wilder books. It wasn't until I read *Prairie Fires* that I placed the arrival of Anna's parents in Minnesota from Norway to the time following the

U.S.-Dakota War. Another inspiration was the novel *Enemy Women* by Paulette Jiles, which I have re-read many times over the past decade. It uses excerpts from Civil War diaries and letters in the chapter headings to provide real-life context to the fictional Adair Colley's quest. (Adair was always on my list of baby names.)

Three recent books set in North Dakota would make great companions to *Windfall*. They include: *The Farmer's Lawyer* by Sarah Vogel, which is her account of the farm crisis of the Reagan years; *Boys and Oil* by Taylor Broby, a poignant memoir of growing up gay in North Dakota; and the novel *O Beautiful* by Jung Yun, a book about a reporter who returns to North Dakota to tell the nitty gritty story of the oil boom in a place she left behind. I also recently loved *Ancestor Trouble* by Maud Newton and the memoir *Lost & Found* by Kathryn Schulz.

NOTES

Work on *Windfall* took place over the span of eight years. During that time, I drew on countless sources—news accounts, obituaries, books, films, photographs, and archives—as well as my own memory, notes, observations, and experiences during ten research trips to North Dakota. I took thousands of my own photographs and videos. I also relied on interviews and research for articles I wrote for *McClatchy*, *E&E News, Washington Post,* and *Stateline* and other outlets. Interviews are noted within the text and below. Most interviews were recorded and transcribed, with several exceptions, which are also noted within the text. Direct quotes from news accounts written by other journalists are attributed to the publication, within the text. I've also included references to interviews, books, photographs, films, or other mediums that, while not always quoted or mentioned directly, served as background information or inspiration for each chapter.

Most conversations with my family are recreated from memory; others are direct quotes from emails and texts or are paraphrased as I recall them.

North Dakota crude oil prices at the start of each chapter are derived from an index compiled by the U.S. Energy Information Administration: https://www.eia.gov/petroleum/.

OTHER WORKS CONSULTED

The following books shaped my thinking about the history and effects of land and mineral booms and busts in North Dakota and beyond—as well as what climate change means to the future of the West:

Anderson, Sam. *Boom Town: The Fantastical Saga of Oklahoma City, Its Chaotic Founding, Its Apocalyptic Weather, Its Purloined Basketball Team, and the Dream of Becoming a World-Class Metropolis.* New York: Crown, 2018.

Briody, Blaire. *The New Wild West: Black Gold, Fracking, and Life in a North Dakota Boomtown.* New York: St. Martin's Press, 2017.

Broom, Sarah M. *The Yellow House.* New York: Grove Press, 2020.

Gold, Russell. *The Boom: How Fracking Ignited the American Energy Revolution and Changed the World.* New York: Simon & Schuster, 2015.

Grandin, Greg. *The End of the Myth.* New York: Henry Holt and Co., 2019.

Griswold, Eliza. *Amity and Prosperity: One Family and the Fracturing of America.* New York: Picador, 2019.

Hansen, Karen V. *Encounter on the Great Plains: Scandinavian Settlers and the Dispossession of Dakota Indians, 1890–1930.* New York: Oxford University Press, 2016.

Murdoch, Sierra Crane. *Yellow Bird: Oil, Murder, and a Woman's Search for Justice in Indian Country.* New York: Random House, 2020.

Rao, Maya. *Great American Outpost: Dreamers, Mavericks, and the Making of an Oil Frontier.* New York: PublicAffairs, 2018.

Treuer, David. *The Heartbeat of Wounded Knee: Native America from 1890 to the Present.* London: Corsair, 2020.

Van der Voo, Lee. *As the World Burns: The New Generation of Activists and the Landmark Legal Fight against Climate Change.* Portland, OR: Timber Press, 2020.

EPIGRAPH

Patricia Nelson Limerick, *The Legacy of Conquest* (New York: W.W. Norton & Co., 1987), 56.

CHAPTER 1: FRACTURED

1   *Windfall* began with an email from my mother in 2009 telling us about the mineral rights she inherited.

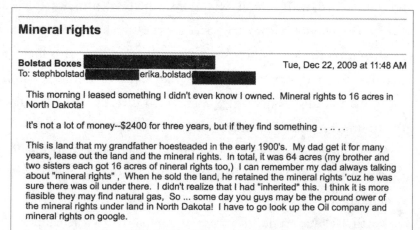

Mineral rights

Bolstad Boxes [redacted]   Tue, Dec 22, 2009 at 11:48 AM
To: stephbolstad [redacted] erika.bolstad [redacted]

This morning I leased something I didn't even know I owned. Mineral rights to 16 acres in North Dakota!

It's not a lot of money--$2400 for three years, but if they find something . . . . . .

This is land that my grandfather hoesteaded in the early 1900's. My dad get it for many years, lease out the land and the mineral rights. In total, it was 64 acres (my brother and two sisters each got 16 acres of nineral rights too,) I can remember my dad always talking about "mineral rights" , When he sold the land, he retained the mineral rights 'cuz he was sure there was oil under there. I didn't realize that I had "inherited" this. I think it is more fiasible they may find natural gas, So ... some day you guys may be the pround ower of the mineral rights under land in North Dakota! I have to go look up the Oil company and mineral rights on google.

Erika, Hope you've been shovel out/melted away from the snow.
Steph, Glad everything is looking great!

Mom

2 "Nancy H. Bolstad Obituary," *Idaho Statesman*, March 7, 2010, www.legacy.com/us/obituaries /idahostatesman/name/nancy-bolstad-obituary?pid=140311112.

CHAPTER 2: PRAIRIE POTHOLES

1 Blake Ellis, "Double Your Salary in the Middle of Nowehere, North Dakota," *CNNMoney*, October 20, 2011, https://www.money.cnn.com/2011/09/28/pf/north_dakota_jobs/index .htm.

2 Stacey Vanek Smith, North Dakota Is Desperate to Find Workers, *Marketplace,* October 12, 2011, https://www.marketplace.org/2011/10/12/north-dakota-desperate-find-workers/.

3 Sean Cockerham and Erika Bolstad, "Obama's Climate Plan Aims Straight at Coal," McClatchy Newspapers, June 25, 2013, https://www.mcclatchydc.com/news/nation-world /national/article24750421.html.

4 Barack Obama, *Inaugural Address by President Barack Obama*, National Archives and Records Administration, Jan. 21, 2013, https://obamawhitehouse.archives.gov/the-press-office/2013/01/21 /inaugural-address-president-barack-obama.

5 Barack Obama, *Remarks by President Obama at the Brandenburg Gate–Berlin, Germany*, National Archives and Records Administration, June 19, 2013, https://www.obamawhitehouse.archives .gov/the-press-office/2013/06/19/remarks-president-obama-brandenburg-gate-berlin germany.

6 "Hydraulic Fracturing." U.S. Geological Survey, https://www.usgs.gov/mission-areas/water -resources/science/hydraulic-fracturing#overview.

7 Christina Nunez, "How Has Fracking Changed Our Future?" *National Geographic*, March 2013, https://www.nationalgeographic.com/environment/article/how-has-fracking-changed-our-future.

8 Barack Obama, *Remarks by the President at U.N. Climate Change Summit*, National Archives and Records Administration, September 23, 2014, https://obamawhitehouse.archives.gov /the-press-office/2014/09/23/remarks-president-un climate-change-summit.

9 Barack Obama, *Remarks by the President on Climate Change*, National Archives and Records Administration, June 25, 2013, https://www.obamawhitehouse.archives.gov/the-press-office /2013/06/25/remarks-president-climate-change.

10 Michael Levi, "Climate consequences of natural gas as a bridge fuel," *Climatic Change* 118, (2013) 609–623, https://doi.org/10.1007/s10584-012-0658-3.

11 Lenny Bernstein, "Sally Jewell at a Different Kind of Summit: Head of the Department of the Interior," *Washington Post,* December 25, 2013, https://www.washingtonpost.com/lifestyle /style/sally-jewell-at-a-different-kind-of-summit-head-of-the-department-of-the-interior/2013 /12/25/ec42aa32 5613-11e3–835d-e7173847c7cc_story.html.

12 "Lostwood National Wildlife Refuge, Annual Narrative Report," U.S. Fish and Wildlife Service, 1987, 22. https://ecos.fws.gov/ServCat/DownloadFile/40699.

13 "Lostwood National Wildlife Refuge," U.S. Fish and Wildlife Service, https://www.fws.gov /refuge/lostwood.

14 "Natural Gas Flaring and Venting: State and Federal Regulatory Overview, Trends, and Impacts," U.S. Department of Energy, 2019, https://www.energy.gov/fecm/state-level-natural -gas-flaring-and-venting-regulations.

15 C. Albert White, "A History of the Rectangular Survey System," U.S. Department of the Interior, Bureau of Land Management, 1983.

CHAPTER 3: RIGGED

1 Bill Shemorry, *Mud, Sweat and Oil: The Early Years of the Williston Basin.* (B. Shemorry: 1991), 53–113.

2 John P. Bluemle, "The 50th Anniversary of the Discovery of Oil in North Dakota," North Dakota Geological Survey, 2001, www.dmr.nd.gov/ndgs/documents/Publication_List/pdf /MiscSeries/MS-89.pdf.

3 "Oil: The Great Hunter," *Time*, December 1, 1952, content.time.com/time/subscriber/article /0,33009,817494,00.html.

4 William E. (Bill) Shemorry, "Clarence Iverson Discovery Oil Well, " 1951, State Historical Society of North Dakota, https://www.history.nd.gov/archives/manuscripts/inventory/10958.html.

5 "Oil Rush in North Dakota," *Collier's*, October 13, 1951, 54–55.

6   Daniel Yergin, *The Prize: The Epic Quest for Oil, Money and Power* (New York: Free Press, 1991), 392.

7   "Oil: The Great Hunter," *Time*, December 1, 1952, content.time.com/time/subscriber/article /0,33009,817494,00.html.

8   "Oil: The Great Hunter," *Time*, December 1, 1952, content.time.com/time/subscriber/article /0,33009,817494,00.html.

9   "Texas: King of the Wildcatters," *Time*, February 13, 1950, content.time.com/time/subscriber /article/0,33009,811889,00.html.

10   *Boom Town,* directed by Jack Conway, Metro-Goldwyn-Mayer Corp., 1940.

11   Ida Tarbell, *All in the Day's Work: An Autobiography* (New York: Macmillan, 1939), 230.

12   "Health and Safety Risks for Workers Involved in Manual Tank Gauging and Sampling at Oil and Gas Extraction Sites," National Institute for Occupational Safety and Health and the Occupational Safety and Health Administration, February 4, 2016, https://www.cdc.gov /niosh/docs/2016-108/default.html.

13   Stephanie Goldberg, "North Dakota 'an exceptionally dangerous' place to work," *Business Insurance,* https://www.businessinsurance.com/article/00010101/NEWS08/150439982/.

14   Steven Mufson, "For oil driller Harold Hamm, Bakken boom brings more billions and a chance to dabble in politics," *Washington Post*, August 12, 2012, https://www.washingtonpost .com/lifestyle/style/for-oil-driller-harold-hamm-bakken-boom-brings-more-billions-and-a -chance-to-dabble-in-politics/2012/08/12/3906740a-e227-11e1-ae7f-d2a13e249eb2_story.html.

15   Christopher Helman, "America's Oil and Gas Billionaires," *Forbes*, March 4, 2013, www .forbes.com/sites/christopherhelman/2013/03/04/americas-oil-and-gas-billionaires.

16   Clare O'Connor, "Oil Billionaire Harold Hamm's Divorce Could Be World's Most Expensive at over $5 Billion, " *Forbes*, March 26, 2013, www.forbes.com/sites/clareoconnor/2013/03/23 /oil-billionaire-harold-hamms-divorce-could-be-worlds-most-expensive-at-over-5-billion /?sh=184e4e424267.

17   Steven Mufson, "A driller's unshakeable faith: Harold Hamm is an unabashed champion for the oil economy that made him rich," *Washington Post*, August 12, 2012, https://www .washingtonpost.com/blogs/keystone-down-the-line/wp/2012/08/12/a-drillers-unshakeable -faith-harold-hamm-is-an-unabashed-champion-for-the-oil-economy-that-made-him-rich/.

18   Bryan Gruley, "The Man Who Bought North Dakota," *Bloomberg*, January 19, 2012, www .bloomberg.com/news/articles/2012-01-19/the-man-who-bought-north-dakota.

19   Amy Dalrymple, "Dunn County Residents Submit Petition for Grand Jury to Investigate Governor," *Inforum*, November 1, 2012, www.inforum.com/news/dunn-county-residents-submit -petition-grand-jury-investigate-governor-1.

20   "In N.D. Politics, Most Oil Industry Contributions Go to Republicans," *Grand Forks Herald*, July 16, 2012, www.grandforksherald.com/news/2178342-nd-politics-most-oil-industry -contributions-go-republicans.

21   Randy Krehbiel, "Oklahoma oilman Harold Hamm will advise presidential candidate Romney, campaign on energy," *Tulsa World*, March 2, 2012, https://tulsaworld.com/news/local /govt-and-politics/oklahoma-oilman-harold-hamm-will-advise-presidential-candidate-romney -campaign-on-energy/image_98ca2084-e452-57a6-a317-798ea01dc40a.html.

22   Amy Harder, "Harold Hamm on Oil, Climate Change, and His Divorce," *National Journal*, August 19, 2013, https://www.nationaljournal.com/s/72090/harold-hamm-oil-climate-change-his-divorce.

23   Amy Harder, "Interior Secretary Shows Her Energy Savvy in North Dakota," *National Journal*, August 11, 2013, https://www.nationaljournal.com/s/73193/interior-secretary-shows-her-energy -savvy-north-dakota.

24   Michael Carlowicz, "Out of the Blue and Into the Black," *NASA Earth Observatory*, December 5, 2012, https://earthobservatory.nasa.gov/features/IntotheBlack.

25   John Hult, "*Image of the Week–Night Lights in North Dakota*," U.S. Geological Survey, August 27, 2019, www.usgs.gov/media/videos/image-week-night-lights-north-dakota.

26   Robert Krulwich, "A Mysterious Patch of Light Shows up in the North Dakota Dark," *NPR*, January 16, 2013, www.npr.org/sections/krulwich/2013/01/16/169511949/a-mysterious-patch -of-light-shows-up-in-the-north-dakota-dark.

27 Jeff Brady, "Much of North Dakota's Natural Gas Is Going up in Flames," *NPR*, January 30, 2014, www.npr.org/2014/01/30/265396179/much-of-north-dakota-s-natural-gas-is-going-up-in-flames.

28 Chris J. Zygarlicke et al., *Bakken Flares and Satellite Images*, University of North Dakota, 2015, 1–4.

29 Brad Plumer, "The Little-Known Story of How 'Fracking' Entered Our Vocabulary," *Vox*, April 28, 2016, https://www.vox.com/2016/4/28/11521618/where-word-fracking-comes-from.

30 Nick Smith, "Continental CEO is to visit Crosby," *Williston Herald*, October 22, 2011, https://www.willistonherald.com/news/continental-ceo-is-to-visit-crosby/article_e4b60de4-3192-592c-b937-36723cf1b437.html.

31 Christopher Helman, "Tycoon Says North Dakota Oil Field Will Yield 24 Billion Barrels, Among World's Biggest," *Forbes*, October 10, 2012, www.forbes.com/sites/christopherhelman/2011/06/27/tycoon-says-north-dakota-oil-field-will-yield-24-billion-barrels-among-worlds-biggest/?sh=5b012f24678f.

32 Sally Jewell, phone interview, March 24, 2021.

33 Amy Dalrymple, "Interior Secretary Gets Firsthand Look at Bakken," *Grand Forks Herald*, August 7, 2013, www.grandforksherald.com/news/interior-secretary-gets-firsthand-look-bakken-0.

34 Amy Harder, "Harold Hamm on Oil, Climate Change, and His Divorce," *National Journal*, August 19, 2013, https://www.nationaljournal.com/s/72090/harold-hamm-oil-climate-change-his-divorce.

35 Clifford Krauss, "Oil Companies Are Sued for Waste of Natural Gas," *New York Times*, October 17, 2013, https://www.nytimes.com/2013/10/18/business/energy-environment/oil-companies-are-sued-over-natural-gas-flaring-in-north-dakota.html.

36 "Importance of Methane," Global Methane Initiative, U.S. Environmental Protection Agency, https://www.epa.gov/gmi/importance-methane.

37 "Flaring up: North Dakota Natural Gas Flaring More than Doubles in Two Years," *Ceres*, July 29, 2013, www.ceres.org/resources/reports/flaring-north-dakota-natural-gas-flaring-more-doubles-two-years.

38 Jeff Brady, "Much of North Dakota's Natural Gas Is Going up in Flames," *NPR*, January 30, 2014, www.npr.org/2014/01/30/265396179/much-of-north-dakota-s-natural-gas-is-going-up-in-flames.

39 Edith Allison and Ben Mandler, "Air Quality Impacts of Oil and Gas," *American Geosciences Institute*, January 28, 2020, https://www.americangeosciences.org/critical-issues/petroleum-environment.

40 "Secretary Jewell Visits North Dakota's Bakken Region," U.S. Department of the Interior, August 7, 2013, www.doi.gov/news/pressreleases/secretary-jewell-visits-north-dakotas-bakken-region.

41 Amy Harder, "My Week in Oil Boom Country," *National Journal*, August 11, 2013, https://news.yahoo.com/week-oil-boom-country-135359586.html.

42 Amy Harder, "Meet the Billionaire Pioneer of America's Oil Boom," *National Journal*, August 19, 2013, https://www.nationaljournal.com/s/72038/meet-billionaire-pioneer-americas-oil-boom.

CHAPTER 4: THE BOOM

1 Elwyn B. Robinson, *History of North Dakota* (Lincoln: University of Nebraska Press, 1968), 398.

2 Russell Lee, "Hardware store window, Crosby, North Dakota," 1937, Library of Congress, www.loc.gov/collections/fsa-owi-black-and-white-negatives/about-this-collection.

3 "East Helena Site Profile," Environmental Protection Agency, October 20, 2017, https://cumulis.epa.gov/supercpad/cursites/csitinfo.cfm?id=0800377.

4 *Helena Independent-Record*, Sept. 24, 1945, https://www.newspapers.com.

5 *Montana Standard*, May 6, 1946, https://www.newspapers.com.

6 "National Register of Historic Places Inventory," U.S. Department of the Interior, 1985, https://npgallery.nps.gov/NRHP/GetAsset/NRHP/64000474_text.

7 Chip Brown, "North Dakota Went Boom," *New York Times*, Jan. 31, 2013, www.nytimes.com/2013/02/03/magazine/north-dakota-went-boom.html.

8 "Split Estate Leaves People Living Near Mineral Deposits at the Mercy of Energy Companies," *Western Organization of Resource Councils*, July 12, 2021, www.worc.org/split-estate-leaves-home-and-landowners-at-the-mercy-of-energy-companies.

9   Colin Jerolmack, *Up to Heaven and Down to Hell* (Princeton: Princeton University Press, 2021), 16.
10   *Gasland,* directed by Josh Fox (HBO, 2010).
11   Neal Conan, "Sparks Fly over 'Gasland' Drilling Documentary," *NPR*, February 24, 2011, www
     .npr.org/2011/02/24/134031183/Gasland-Takes-On-Natural-Gas-Drilling-Industry.
12   "Code of Ethics and Standards of Practice," American Association of Petroleum Landmen,
     https://www.landman.org/about/governance/code-of-ethics-and-standards-of-practice.
13   "Walter Braun Obituary," *Bozeman Daily Chronicle*, April 4, 2014, www.legacy.com/obituaries
     /bozemandailychronicle/obituary-print.aspx?n=walter-l-braun&pid=170481322.
14   "Oil Rush in North Dakota," *Collier's*, October 13, 1951, 54–55.
15   Pamela Reisel, "Life After the Movies: Former Film Stars Transition into 1950s Television," Film
     in American Popular Culture, June 2006, https://www.americanpopularculture.com/archive
     /film/former_film_stars.htm.
16   *Louisiana Story*, directed by Robert Flaherty (Lopert Films Inc., 1948).
17   *American Frontier*, directed by Willard Van Dyke (American Petroleum Institute, 1953).

CHAPTER 5: PAST IS PROLOGUE

1   "Homestead Act: Primary Documents in American History: Introduction," Library of Congress,
     https://guides.loc.gov/homestead-act.
2   "Sustainability Plan," U.S. National Archives and Records Administration, September 2021,
     https://www.archives.gov/about/plans-reports/sustainability.
3   "Land Entry Case Files and Related Records," U.S. National Archives and Records
     Administration, https://www.archives.gov/research/land/land-records.
4   "Research in the Land Entry Files of the General Land Office," U.S. National Archives and
     Records Administration, 2009, https://www.archives.gov/files/publications/ref-info-papers
     /rip114.pdf.
5   Gregory Paynter Shine, "The War and Westward Expansion," National Park Service, U.S.
     Department of the Interior, https://www.nps.gov/articles/the-war-and-westward-expansion.htm.
6   Todd Arrington, "Abraham Lincoln and the West," National Park Service, U.S. Department
     of the Interior, https://www.nps.gov/articles/abraham-lincoln-and-the-west.htm.
7   Trina Williams Shanks, "The Homestead Act: A Major Asset-Building Policy in American
     History," (St. Louis, MO: Washington University, Center for Social Development, 2000),
     https://doi.org/10.7936/K7RR1XQM.
8   Richard Edwards, "The Disappearing Story of the Black Homesteaders Who Pioneered
     the West," *Washington Post*, July 5, 2018, https://www.washingtonpost.com/opinions/the
     -disappearing-story-of-the-black-homesteaders-who-pioneered-the-west/2018/07/05/ca0b51b6
     -7f09-11e8-b0ef-fffcabeff946_story.html.
9   "African American Homesteaders in the Great Plains," National Park Service, U.S. Department
     of the Interior, https://www.nps.gov/articles/the-war-and-westward-expansion.htm. https://
     www.nps.gov/articles/african-american-homesteaders-in-the-great-plains.htm.
10   "Today in History—June 2—Indian Citizenship Act," Library of Congress, https://www.loc
     .gov/item/today-in-history/june-02/.
11   David Miller et al., *The History of the Assiniboine and Sioux Tribes of the Fort Peck Indian Reservation,
     Montana, 1600–2012* (Poplar, MT: Fort Peck Community College, 2012), 113–154.
12   "Homesteading by the Numbers," National Park Service, U.S. Department of the Interior,
     https://www.nps.gov/home/learn/historyculture/bynumbers.htm.
13   John C. Hudson, *Plains Country Towns* (Minneapolis: University of Minnesota Press, 1985), 62.
14   Claire Strom, *Profiting from the Plains: The Great Northern Railway and Corporate Development of the
     American West* (Seattle: University of Washington Press, 2003), 83–85.
15   *Ward County Independent,* February 1, 1906, https://chroniclingamerica.loc.gov/lccn/sn88076421
     /1906-02-01/ed-1/seq-3/.
16   *Ward County Independent*, December 27, 1905, https://chroniclingamerica.loc.gov/lccn/sn88076421
     /1905-12-27/ed-2/seq-2/.
17   *Ward County Independent*, December 13, 1905, https://chroniclingamerica.loc.gov/lccn/sn88076421
     /1905-12-13/ed-1/seq-2/.

18 Elwyn B. Robinson, *History of North Dakota* (Lincoln: University of Nebraska Press, 1968), 317.

19 *Ward County Independent*, December 27, 1905, https://chroniclingamerica.loc.gov/lccn/sn88076421
/1905-12-27/ed-1/seq-2/.

20 *Ward County Independent*, December 27, 1905, https://chroniclingamerica.loc.gov/lccn/sn88076421
/1905-12-27/ed-1/seq-2/.

21 *Ward County Independent*, February 14, 1907, https://chroniclingamerica.loc.gov/lccn/sn88076421
/1907-02-14/ed-1/seq-7/.

22 H. Elaine Lindgren, *Land in Her Own Name: Women as Homesteaders in North Dakota* (Norman,
OK: University of Oklahoma Press, 1996), 175.

23 Lindgren, 140.

24 Lindgren, 177.

25 Lindgren, 176.

26 Anna J. Sletvold homestead file, Records of the Bureau of Land Management, National Archives
Building, Washington, DC.

CHAPTER 6: MANIFEST DESTINY

1 Anna J. Sletvold homestead file, Records of the Bureau of Land Management, National Archives
Building, Washington, DC.

2 "Fire of 1893," *Fargo, North Dakota: Its History and Images*, North Dakota State University Archives,
https://library.ndsu.edu/fargo-history/?q=content/fire-1893.

3 Karen Herzog, "North Dakota's heritage of alcohol," *Bismarck Tribune*, Dec 9, 2012, https://
bismarcktribune.com/news/state-and-regional/north-dakota-s-heritage-of-alcohol/article
_6fd6cc22-3fca-11e2-ad03-001a4bcf887a.html.

4 Kate Savageau, "Moorhead's Saloon Era, 1890–1915," MNopedia, Minnesota Historical Society,
https://www.mnopedia.org/event/moorhead-s-saloon-era-1890-1915.

5 *Fargo Forum and Daily Republican*, November 22, 1905, https://chroniclingamerica.loc.gov
/lccn/sn85042224/1905-11-22/ed-1/seq-11/.

6 *Fargo Forum and Daily Republican*, July 6, 1904, https://chroniclingamerica.loc.gov/lccn
/sn85042224/1904-07-06/ed-1/seq-12/.

7 "Swami Vivekananda and His 1893 Speech," Art Institute of Chicago, https://www.artic.edu
/swami-vivekananda-and-his-1893-speech.

8 "Love, Not Gold, Sways the Divine Powers of the Great Hindoo," *Fargo Forum and Daily Republican*,
March 2, 1904, https://chroniclingamerica.loc.gov/lccn/sn85042224/1904-03-02/ed-1/seq-8/.

9 "A Busy Man," *Fargo Forum and Daily Republican*, February 22, 1904, https://chroniclingamerica
.loc.gov/lccn/sn85042224/1904-02-22/ed-1/seq-4/.

10 "The Prince in Fargo," *Fargo Forum and Daily Republican*, February 13, 1904, https://
chroniclingamerica.loc.gov/lccn/sn85042224/1904-02-13/ed-1/seq-5/.

11 Robinson, *History of North Dakota*, 242.

12 *Fargo Forum and Daily Republican*, January 2, 1904, https://chroniclingamerica.loc.gov/lccn
/sn85042224/1904-01-02/ed-1/seq-7/.

13 *Fargo Forum and Daily Republican*, January 6, 1904, https://chroniclingamerica.loc.gov/lccn
/sn85042224/1904-01-06/ed-1/seq-11/.

14 *Fargo Forum and Daily Republican*, June 15, 1905, https://chroniclingamerica.loc.gov/lccn
/sn85042224/1905-06-15/ed-1/seq-7/.

15 Andrew Haraseth homestead file, Records of the Bureau of Land Management, National
Archives Building, Washington, DC.

16 Gilbert C. Fite, *The Farmer's Frontier: 1865–1900* (Albuquerque, NM: University of New Mexico
Press, 1977) 215–217.

17 *Ward County Independent*, December 13, 1905, https://chroniclingamerica.loc.gov/lccn/sn88076421
/1905-12-13/ed-1/seq-2/.

18 *Fargo Forum and Daily Republican*, January 26, 1904, https://chroniclingamerica.loc.gov/lccn
/sn85042224/1904-01-26/ed-1/seq-4/.

19 *Report of the Public Lands Commission* (U.S. Government Printing Office: 1905) 73–77, https://
hdl.handle.net/2027/hvd.32044032089740.

20   Edwards, Richard, et al. *Homesteading the Plains: Toward a New History* (Lincoln : University of Nebraska Press, 2017), 60.

CHAPTER 7: MOTHERLESS

1   John W. Mason, *History of Otter Tail County: Its People, Industries and Institutions* (Indianapolis, IN: B.F. Bowen, 1916), preface.
2   Mason, 182.
3   Samuel Meshbesher, "Norwegian Immigration in Minnesota," MNopedia, Minnesota Historical Society, http://www.mnopedia.org/norwegian-immigration-minnesota.
4   Caroline Fraser, *Prairie Fires: The American Dreams of Laura Ingalls Wilder* (New York: Metropolitan Books, 2017), 10–24.
5   "The U.S.-Dakota War of 1862: Aftermath," Minnesota Historical Society, https://www.usdakotawar.org/history/aftermath.
6   Mason, *History of Otter Tail County: Its People, Industries and Institutions,* 550–551.
7   The Panic of 1873, PBS, https://www.pbs.org/wgbh/americanexperience/features/grant-panic/.
8   *Fergus Falls Weekly Journal,* January–March 1887, Otter Tail County Historical Society, https://www.otchs.org/.
9   *Fergus Falls Weekly Journal,* January 20, 1887, Otter Tail County Historical Society, https://www.otchs.org/.
10 *Fergus Falls Ugeblad,* May 11, 1887, https://chroniclingamerica.loc.gov/lccn/sn83025227/1887-05-11/ed-1/seq-4/.
11 *Fergus Falls Ugeblad,* May 11, 1887.
12 David Laskin, *The Children's Blizzard* (New York: Harper Perennial, 2004), 269.
13 *Fergus Falls Ugeblad,* Sept 7, 1892, https://chroniclingamerica.loc.gov/lccn/sn83025227/1892-09-07/ed-1/seq-8/.

CHAPTER 8: ANNA

1   Erika Bolstad, "In North Dakota, Searching for Her Roots and Finding a Lot to Root For," *Washington Post,* March 29, 2014, www.washingtonpost.com/lifestyle/travel/in-north-dakota-searching-for-her-roots-and-finding-a-lot-to-root-for/2014/03/27/8be108ee-b056-11e3-9627-c65021d6d572_story.html.
2   Medical records retrieved by the author in 2013 at the North Dakota State Hospital.
3   "Biennial Report of the North Dakota State Hospital, 1902–1904," North Dakota State Documents, North Dakota State Library.
4   "Biennial Report of the North Dakota State Hospital, 1918–1920," North Dakota State Documents, North Dakota State Library.
5   James Smorada, *Century of Stories: Jamestown and Stutsman County* (Fort Seward Historical Society: 1983), 85–90.
6   Mrs. W.F. Cushing, *Glimpses of Jamestown and Stutsman County,* 1915.
7   "Biennial Report of the North Dakota State Hospital, 1908–1910," North Dakota State Documents, North Dakota State Library.
8   "Biennial Report of the North Dakota State Hospital, 1918–1920," North Dakota State Documents, North Dakota State Library.
9   "Biennial Report of the North Dakota State Hospital, 1920–1922," North Dakota State Documents, North Dakota State Library.
10 Irving Berlin, "All By Myself," 1921.

CHAPTER 9: A WOMAN BEHIND EVERY TREE

1   Samantha Grossman, "Highest Rent in United States in Williston, North Dakota, Not New York," *Time,* February 18, 2014, https://time.com/8731/highest-rent-in-us-williston-north-dakota/.
2   Jackson Bolstad, "Clarifying Williston's Ban on RVs," *Williston Herald,* July 14, 2012, www.willistonherald.com/news/clarifying-williston-s-ban-on-rvs/article_cb7971d8-ce14-11e1-9089-001a4bcf887a.html.
3   Lauren Donovan, "Williston Walmart Becomes a Mecca for Oil Job Seekers," *Bismarck Tribune,*

January 29, 2012, bismarcktribune.com/news/state-and-regional/williston-walmart-becomes-a-mecca-for-oil-job-seekers/article_c044ba74-4977-11e1-920b-0019bb2963f4.html.

4 Amy Dalrymple, "Man Who Killed Montana Teacher given 100 Year Sentence," *Bismarck Tribune*, April 17, 2015, bismarcktribune.com/news/state-and-regional/man-who-killed-montana-teacher-given-100-year-sentence/article_3d88cad6-54ac-55a8-95ac-9552a1f7c114.html.

5 "2013 Traffic Fatalities by State," National Highway Traffic Safety Administration, U.S. Department of Transportation, https://cdan.nhtsa.gov/stsi.htm.

6 "Examining Violence Against American Indian and Alaska Native Women," National Institute of Justice, June 16, 2016, https://nij.ojp.gov/topics/articles/examining-violence-against-american-indian-and-alaska-native-women.

7 Jenny Michael, "Woman Found Dead in McLean County," *Bismarck Tribune*, March 11, 2014, https://www.bismarcktribune.com/news/local/crime-and-courts/woman-found-dead-in-mclean-county/article_8d832b06-a966-11e3-8412-0019bb2963f4.html.

8 Corey Henne, "Deborah Parker," *Badass Womxn in the Pacific Northwest*, June 10, 2019, https://uw.pressbooks.pub/badasswomxninthepnw/.

9 "Robin Fox Obituary," *Minot Daily News*, March 12, 2014, www.minotdailynews.com/obituaries/2014/03/robin-fox/.

10 Sierra Crane Murdoch, "On the Trail of a Murder in the Dakota Badlands," *Literary Hub*, February 25, 2020, lithub.com/on-the-trail-of-a-murder-in-the-dakota-badlands/.

11 Phil Davies, "Homeland of Opportunity, " Federal Reserve Bank of Minneapolis, October 6, 2014, https://www.minneapolisfed.org/article/2014/homeland-of-opportunity.

12 Rick Ruddell, Dheeshana S. Jayasundara, Roni Mayzer, and Thomasine Heitkamp, "Drilling Down: An Examination of the Boom-Crime Relationship in Resource Based Boom Counties," *Western Criminology Review*, 15, No. 1 (2014): 3–17, https://www.westerncriminology.org/documents/WCR/v15n1/v15n1.pdf

13 Jenna Ebersole, "Number of Sex Offenders on the Rise," *Williston Herald*, Oct 30, 2012, https://www.willistonherald.com/news/number-of-sex-offenders-on-the-rise/article_181643ca-22a4-11e2-a27c-0019bb2963f4.html.

14 *The Overnighters,* directed by Jesse Moss (Drafthouse Films, 2014.)

15 John Eligon, "An Oil Town Where Men Are Many, and Women Are Hounded," *New York Times*, January 16, 2013, https://www.nytimes.com/2013/01/16/us/16women.html.

16 Dave Kornder, interview with the author, February 27, 2014.

17 David Shaffer and Evan Ramstad, "NTSB: 400,000 Gallons of Crude Spilled in N.D. Train Wreck," *Star Tribune*, January 23, 2014, www.startribune.com/ntsb-400-000-gallons-of-crude-spilled-in-casselton-train-wreck/239948631/.

18 Jessica Murphy, "Lac-Megantic: The Runaway Train That Destroyed a Town," BBC, January 19, 2018, www.bbc.com/news/world-us-canada-42548824.

19 Alicia Underlee Nelson, "North Dakota Republican Calls for Oil Boom Slowdown after Train Crash," *Reuters*, January 3, 2014, www.reuters.com/article/northdakota-train-politician/north-dakota-republican-calls-for-oil-boom-slowdown-after-train-crash-idUSL2N0KD1DQ20140103.

20 Amy Dalrymple, "Hamm Fires Back after Call to Slow Oil Development," *Dickinson Press*, January 9, 2014, www.thedickinsonpress.com/business/energy-and-mining/2044835-hamm-fires-back-after-call-to-slow-oil-development.

21 Robert Harms, "We Should Be Able to Have a Thoughtful Discussion About Oil Safety," *Say Anything*, January 6, 2014, www.sayanythingblog.com/entry/harms-column-we-should-be-able-to-have-a-thoughtful-discussion-about-oil-safety/.

22 Rob Port, "NDGOP Chairman Apologizes to Party Leadership, Elected Leaders for Oil Regulation Comments," *Say Anything*, January 6, 2014, https://www.sayanythingblog.com/entry/ndgop-chairman-apologizes-to-party-leadership-elected-leaders-for-oil-regulation-comments/.

23 Stephen Fee, "Crude Oil Brings Boom Times and Safety Concerns to North Dakota," *PBS NewsHour*, May 11 2014, www.pbs.org/newshour/show/crude-oil-brings-boom-times-safety-concerns-north-dakota.

24 James MacPherson, "5 Years and $100m Later, Huge North Dakota Oil Spill Finally Cleaned

Up," *Insurance Journal*, September 21, 2018, https://www.insurancejournal.com/news/midwest/2018/09/21/502039.htm.

25 Lauren Donovan, "Radioactive Dump Site Found in Remote North Dakota Town," *Bismarck Tribune*, March 11, 2014, https://www.bismarcktribune.com/bakken/radioactive-dump-site-found-in-remote-north-dakota-town/article_39d0d08a-a948-11e3-8a3b-001a4bcf887a.html.

26 Jeff McMahon, "Strange Byproduct of Fracking Boom: Radioactive Socks," *Forbes*, July 24, 2013, www.forbes.com/sites/jeffmcmahon/2013/07/24/strange-byproduct-of-fracking-boom-radioactive-socks/?sh=c3d60994fa75.

27 Phoebe Tollefson and Shane Sanderson, "Belgrade Man Hit with Fine for Dumping Radioactive Waste He Was Hired to Manage," *Billings Gazette*, October 22, 2020, https://www.billingsgazette.com/news/state-and-regional/crime-and-courts/belgrade-man-hit-with-fine-for-dumping-radioactive-waste-he-was-hired-to-manage/article_472d8e4d-5ee3-5f9c-98e7-6828fe817592.html.

28 Sarah Jane Keller, "North Dakota Wrestles with Radioactive Oilfield Waste, " *High Country News*, July 14, 2014, www.hcn.org/articles/north-dakota-wrestles-with-radioactive-oilfield-waste.

29 Emily Guerin, "In North Dakota, Signs of Changing Attitudes towards Oil and Gas Development," *High Country News*, May 27, 2014, www.hcn.org/blogs/goat/in-north-dakota-signs-of-changing-attitudes-towards-oil-and-gas-development.

30 Valerie Naylor, "Why I'm Speaking out to Protect the Park I Love," National Parks Conservation Association, April 25, 2017, www.npca.org/articles/1526-why-i-m-speaking-out-to-protect-the-park-i-love.

31 Nick Smith, "Plan to Protect 'Extraordinary Places' Unveiled," *Bismarck Tribune*, December 19, 2013, https://www.bismarcktribune.com/bakken/plan-to-protect-extraordinary-places-unveiled/article_447c7bfe-68fe-11e3-bb7c-001a4bcf887a.html.

32 "NDFB Opposes List of Extraordinary Places Proposed by Stenehjem," *Agweek*, February 21, 2014, www.agweek.com/news/3792022-ndfb-opposes-list-extraordinary-places-proposed-stenehjem.

33 Emily Guerin, "A tale of two parks: How the Bakken boom transformed a landscape, " *High Country News*, November 27, 2017, https://www.hcn.org/issues/49.20/public-lands-a-tale-of-two-parks-how-the-bakken-boom-transformed-a-landscape.

34 "BCA Accomplishments for 2014," Badlands Conservation Alliance, May 3, 2016, https://www.badlandsconservationalliance.org/news/36accomplish.

35 Stephanie Gaswirth and Kristen Marra, interview with the author, February 27, 2014.

36 2008 U.S. Geological Survey assessment, https://doi.org/10.3133/fs20083092.

37 Bethany McLean, *Saudi America: The Truth about Fracking and How It's Changing the World* (New York: Columbia Global Reports, 2018), 91.

38 2013 U.S. Geological Survey assessment, https://pubs.usgs.gov/fs/2013/3013/.

39 Amy Dalrymple, "Report More than Doubles Recoverable Oil Estimate for Williston Basin," *Dickinson Press*, April 30, 2013, www.thedickinsonpress.com/business/1824916-report-more-doubles-recoverable-oil-estimate-williston-basin.

40 Adam Wilmoth, "North Dakota, Montana's Oil Potential Revised Even Higher," *Oklahoman*, May 1, 2013, www.oklahoman.com/article/3805035/north-dakota-montanas-oil-potential-revised-even-higher.

CHAPTER 10: LONG UNDERWEAR

1 Erika Bolstad, "Climate Regs Won't Crimp Energy Boom, McCarthy Tells Worried North Dakotans," *E&E News*, March 3, 2014, https://subscriber.politicopro.com/article/eenews/2014/03/03/climate-regs-wont-crimp-energy-boom-mccarthy-tells-worried-north-dakotans-116729.

2 David Laskin, *The Children's Blizzard* (New York: Harper Perennial, 2004), 271.

3 Ian Vestal, interview with the author, March 3, 2014.

4 Erika Bolstad, "Bakken Boom Town Invests in 'Quality of Life,' Builds Palatial Rec Center," *E&E News*, April 7, 2014, https://subscriber.politicopro.com/article/eenews/2014/04/07/bakken-boom-town-invests-in-quality-of-life-builds-palatial-rec-center-114756.

5 Darin Krueger, interview with the author, March 1, 2014.

6 *American Frontier*, directed by Willard Van Dyke (American Petroleum Institute, 1953).

7   Lindgren, *Land in Her Own Name: Women as Homesteaders in North Dakota,* 128–129.
8   Scott Randolph, phone interview with the author, September 16, 2013.

CHAPTER 11: THE MYSTERY BABY-MAKING BOULDER OF LIBERTY, ILLINOIS

1   "Liberty, Illinois: Mystery Baby-Making Boulder," https://www.roadsideamerica.com/tip/418.
2   Stephen E. Ambrose, *Undaunted Courage* (New York: Simon & Schuster, 1996).
3   Elwyn B. Robinson, *History of North Dakota* (Lincoln: University of Nebraska Press, 1968), 41.
4   "Sacagawea statue unveiling," *Jamestown Weekly Alert,* October 20, 1910.
5   Stephen E. Ambrose, *Undaunted Courage,* read by Barrett Whitener (New York: Simon & Schuster, 1996), audiobook, 21 hr., 40 min.
6   Jacqueline Keeler, *Standoff: Standing Rock, the Bundy Movement, and the American Story of Sacred Lands* (Salt Lake City, UT: Torrey House Press, 2021), 159–160.
7   Best For Last Club, https://www.fargomoorhead.org/plan-a-trip/best-for-last-club/.
8   "Eugene Field," Uncommon Character—St. Joseph, Missouri, September 5, 2019, https://uncommoncharacter.com/stories/eugene-field/.
9   "What Happened in Ferguson?" *New York Times,* August 13, 2014, www.nytimes.com/interactive/2014/08/13/us/ferguson-missouri-town-under-siege-after-police-shooting.html.
10 *The Journals of Lewis and Clark,* April 16, 1804, https://lewisandclarkjournals.unl.edu/item/lc.jrn.1804-04-16.

CHAPTER 12: BEATRICE

1   Hugh Jackson Dobbs, *History of Gage County, Nebraska* (Lincoln, NE : Western Publishing and Engraving Company, 1918), 123–127.
2   "Daniel Freeman Homestead," National Park Service, U.S. Department of the Interior, https://www.nps.gov/places/daniel-freeman-homestead.htm.
3   "Oscar Micheaux," National Park Service, U.S. Department of the Interior, https://www.nps.gov/people/oscar-micheaux.htm.
4   "African American Homesteaders in the Great Plains," National Park Service, U.S. Department of the Interior, https://www.nps.gov/articles/african-american-homesteaders-in-the-great-plains.htm.
5   Tiffany Camhi, "A racist history shows why Oregon is still so white," *OPB,* June 9, 2020, https://www.opb.org/news/article/oregon-white-history-racist-foundations-black-exclusion-laws/.
6   Mike Konczal, *Freedom from the Market: America's Fight to Liberate Itself from the Grip of the Invisible Hand* (New York: The New Press, 2021), 30–35.
7   Trina Williams Shanks, *"The Homestead Act: A Major Asset-Building Policy in American History,"* (St. Louis, MO: Washington University, Center for Social Development, 2000), https://doi.org/10.7936/K7RR1XQM.

CHAPTER 13: MEDORA

1   *The Bad Lands Cow Boy,* February 7, 1884, https://chroniclingamerica.loc.gov/lccn/sn84024777/1884-02-07/ed-1/seq-1/.
2   Robinson, *History of North Dakota,* 186–193.
3   Robinson, 186.
4   Edmund Morris, *The Rise of Theodore Roosevelt* (New York: Coward, McCann & Geoghegan, 1979), 266.
5   Robinson, *History of North Dakota,* 191.
6   Owen Wister, *The Virginian* (New York: Pocket Books, 2002), 31.
7   David McCullough, *Brave Companions: Portraits in History* (New York: Simon & Schuster, 1992), 69–85.
8   *The Bad Lands Cow Boy,* February 14, 1884.
9   *The Bad Lands Cow Boy,* February 14, 1884.
10 Owen Wister, *Roosevelt, the Story of a Friendship, 1880–1919* (New York: The Macmillan Company, 1930), 51.
11 David McCullough, *Brave Companions: Portraits in History,* 55–68.
12 Patrick Springer, "'Last Great Hunt': The End of the Great Northern Buffalo Herd Happened

at Standing Rock," *Billings Gazette*, October 12, 2019, billingsgazette.com/news/state-and
-regional/last-great-hunt-the-end-of-the-great-northern-buffalo-herd-happened-at-standing
-rock/article_24a155e4-8b11-5b46-a242-0c1aea0737bd.html.

13   William Hornady, *The Extermination of the American Bison: Report of the National Museum, 1886–87* (Washington, DC: Government Printing Office, 1889), 510.

14   Edmund Morris, *The Rise of Theodore Roosevelt* (New York: Coward, McCann & Geoghegan, 1979), 191.

15   Alysa Landry, "Theodore Roosevelt: 'The Only Good Indians Are the Dead Indians,'" *Indian Country Today*, June 28, 2016, indiancountrytoday.com/archive/theodore-roosevelt-the-only
-good-indians-are-the-dead-indians.

16   Morris, 211.

17   Robinson, 193.

18   "Marquis De Morès," National Park Service, U.S. Department of the Interior, https://www
.nps.gov/thro/learn/historyculture/marquis-de-mores.htm.

19   David Gessner, *Leave It As It Is* (New York: Simon & Schuster, 2020), 67.

20   Letter from Theodore Roosevelt to Albert T. Volwiler, Theodore Roosevelt Papers, Library of Congress Manuscript Division, https://www.theodorerooseveltcenter.org/Research/Digital
-Library/Record?libID=o263592.

21   Larry Woiwode, *Aristocrat of the West: The Story of Harold Schafer* (Fargo: Institute for Regional Studies, North Dakota State University, 2000), 311.

22   The author attended the Medora Musical on September 4, 2014; June 19, 2020; August 20, 2020; and June 26, 2021.

23   Emily Walter, phone interview, May 7, 2021.

24   *The One and Only, Genuine, Original Family Band*, directed by Michael O'Herlihy (Disney, 1968).

25   Renata Adler, "Election Comedy Fails to Drum up Support," *New York Times*, March 22, 1968, www
.nytimes.com/1968/03/22/archives/film-one-and-only-genuine-original-family-banddisney
-musical-opens.html.

26   Alysa Landry, "Theodore Roosevelt: 'The Only Good Indians Are the Dead Indians,'" *Indian Country Today*, June 28, 2016, indiancountrytoday.com/archive/theodore-roosevelt-the
-only-good-indians-are-the-dead-indians.

27   Steven S. Schwarzschild, "The Marquis De Mores, the Story of a Failure (1858–1896)," *Jewish Social Studies*, vol. 22, no. 1 (1960): 3–26, www.jstor.org/stable/4465751.

CHAPTER 14: THE BUST

1   Derrick Braaten, interview with the author, September 10, 2014.

2   John Ydstie, "It's Still Too Early for Tanking Oil Prices to Curb U.S. Drilling," *NPR*, November 4, 2014, www.npr.org/2014/11/04/361380522its-still-too-early-for-tanking-oil-prices-to-curb-u-s-drilling.

3   Deborah Sontag, "Where Oil and Politics Mix," *New York Times*, November 24, 2014, www
.nytimes.com/interactive/2014/11/24/us/north-dakota-oil-boom-politics.html.

4   Chris Giles, "Winners and Losers of Oil Price Plunge," *Financial Times*, December 15, 2014, www.ft.com/content/3f5e4914-8490-11e4-ba4f-00144feabdc0#axzz3M0pu7m13.

5   Clifford Krauss, "Despite Slumping Prices, No End in Sight for U.S. Oil Production Boom," *New York Times*, October 17, 2014, www.nytimes.com/2014/10/18/business/energy-environment/us
-oil-boom-shows-no-signs-of-slowing-down.html.

6   Stephen P.A. Brown and Mine K. Yucel, "The Shale Gas and Tight Oil Boom," *Council on Foreign Relations*, October 15, 2013, www.cfr.org/report/shale-gas-and-tight-oil-boom.

7   "If the Bakken is Busting, Just Say So," *Williston Herald*, December 14, 2014, A4. https://issuu
.com/wickcommunications/docs/12-14-14_williston.

8   Manny Fernandez and Jeremy Alford, "Some States See Budgets at Risk as Oil Price Falls," *New York Times*, December 26, 2014, www.nytimes.com/2014/12/27/us/falling-oil-prices-have
-ripple-effect-in-texas-louisiana-oklahoma.html.

9   "It's Still Too Early for Tanking Oil Prices to Curb U.S. Drilling," *NPR*, November 4, 2014, https://www.npr.org/transcripts/361380522.

10   "'The More We Delay, the More We Will Pay,' Says Ban, Urging Action on Climate," *UN News*, December 11, 2014, https://news.un.org/en/story/2014/12/486092.

11 "Budget a Big Step for Oil Patch," *Williston Herald*, December 8, 2014, www.willistonherald.com /opinion/editorials/budget-a-big-step-for-oil-patch/article_53bc5b72-7ef9-11e4-b881 -a322152b7275.html.

12 "Budget a Big Step for Oil Patch," *Williston Herald*, December 8, 2014, www.willistonherald .com/opinion/editorials/budget-a-big-step-for-oil-patch/article_53bc5b72-7ef9-11e4-b881 -a322152b7275.html.

13 Rebecca Fett, *It Starts with the Egg* (New York: Franklin Fox Publishing, 2014), 39–60.

14 *Into the Woods*, directed by Rob Marshall (Disney, 2014).

15 Derrick Braaten, phone interview with the author, January 6, 2015.

16 Rob Grunewald and Dulguun Batbold, "The Bakken Faces a Steep Drop in Oil Prices," *Federal Reserve Bank of Minneapolis*, February 6, 2015, www.minneapolisfed.org/publications/fedgazette /the-bakken-faces-a-steep-drop-in-oil-prices.

17 Christopher Helman, "Billionaire Harold Hamm Slashes 2015 Drilling on Low Oil Prices," *Forbes*, December 23, 2014, www.forbes.com/sites/christopherhelman/2014/12/22/billionaire -harold-hamm-slashes-2015-drilling-on-low-oil-prices/?sh=8e1e6f618bd4.

CHAPTER 16: HYDROCARBONS

1 Julie LeFever, interview with the author, June 13, 2016.

2 Juan Miguel Pedraza, "Up-Close Tour of North Dakota's Underground Geology," *University of North Dakota*, February 2014, www1.und.edu/features/2014/02/core-sample-library.cfm.

3 Jan Orvik, "Core Investment: State Unveils Expanded Core and Sample Library on Campus," *University Letter*, October 4, 2016, blogs.und.edu/uletter/2016/09/core-investment-state-unveils -expanded-core-and-sample-library-on-campus/.

4 John P. Bluemle, "*The 50th Anniversary of the Discovery of Oil in North Dakota*," North Dakota Geological Survey, 2001, www.dmr.nd.gov/ndgs/documents/Publication_List/pdf/MiscSeries /MS-89.pdf.

5 "Julie LeFever Obituary," *Grand Forks Herald*, January 11, 2017, www.grandforksherald.com /obituaries/4195717-julie-lefever.

6 Jan Orvik, "Largely Unknown Legacy of Julie LeFever," *UND Today*, April 4, 2017, blogs.und.edu /und-today/2017/04/largely-unknown-legacy-of-julie-lefever/.

7 "Julie LeFever Obituary," *Grand Forks Herald*, January 11, 2017, www.grandforksherald.com /obituaries/4195717-julie-lefever.

8 Jan Orvik, "Largely Unknown Legacy of Julie LeFever," *UND Today*, April 4, 2017, blogs.und.edu /und-today/2017/04/largely-unknown-legacy-of-julie-lefever/.

9 2014 National Climate Assessment: https://nca2014.globalchange.gov/.

10 2018 National Climate Assessment, Northern Great Plains: https://nca2018.globalchange .gov/chapter/22/.

11 Yale Climate Opinion Maps: https://climatecommunication.yale.edu/visualizations-data/ycom-us/.

12 Kolby KickingWoman, "Dakota Access Pipeline Timeline," *Indian Country Today*, July 9, 2020, https://indiancountrytoday.com/news/dakota-access-pipeline-timeline.

13 Amy Dalrymple, "Trump Ready to 'Get out of the Way' of Oil Industry," *Bemidji Pioneer*, May 26, 2016, www.bemidjipioneer.com/news/4041797-trump-ready-get-out-way-oil-industry.

14 Erika Bolstad, "'We Got Our Butt Kicked' by Flooding, and Are Adapting," *E&E News*, September 15, 2016, https://subscriber.politicopro.com/article/eenews/2016/09/15/we-got-our -butt-kicked-by-flooding-and-are-adapting-070614.

15 Erika Bolstad, "North Dakota Adapts to Climate Change without Saying It's Real," *E&E News*, September 14, 2016, https://www.eenews.net/articles/n-d-adapts-to-climate-change -without-saying-its-real/.

16 Robert Nordstrom, "New Terminal at Minot Int'l Dramatically Increases Capacity & Prepares Community for Future," *Airport Improvement*, May–June 2016, airportimprovement.com/article /new-terminal-minot-intl-dramatically-increases-capacity-prepares-community-future.

CHAPTER 17: SANS SOUCI

1 Erika Bolstad, "La. governor touts resilience plans, bids to diversify economy," *E&E News*,

March 7, 2017, https://subscriber.politicopro.com/article/eenews/2017/03/07/la-governor-touts
-resilience-plans-bids-to-diversify-economy-062589.

2   Erika Bolstad, "Trump Country is Flooding, and Climate Ideas Are Shifting," *E&E News*,
May 15, 2017, https://www.eenews.net/articles/trump-country-is-flooding-and-climate-ideas
-are-shifting/.

3   Joan Didion, *South and West* (New York: Alfred A. Knopf, 2017), 19.

4   Didion, 26.

5   Didion, 55.

6   Didion, 31.

7   Erika Bolstad, "Trump Country is Flooding, and Climate Ideas Are Shifting," *E&E News*,
May 15, 2017, https://www.eenews.net/articles/trump-country-is-flooding-and-climate-ideas
-are-shifting/.

8   Didion, 44.

9   "Levee," City of Cape Girardeau, https://www.cityofcapegirardeau.org/departments/public
_works/stormwater/Levee.

10 Mark Twain, *Life on the Mississippi* (Boston: James R. Osgood & Co., 1883), 130.

CHAPTER 18: THE ANTHROPOCENE

1   Paul Cartwright, "How Last Chance Gulch Got Its Name," *Helena Independent Record*, August 5,
2016, helenair.com/news/opinion/guest/how-last-chance-gulch-got-its-name/article_7596bd3d-54e5
-579c-8ca4-6147b43fbe5b.html.

2   Jim Robbins, "Spiffing up Helena's Last Chance Gulch," *New York Times*, August 2, 1998,
www.nytimes.com/1998/08/02/realestate/spiffing-up-helena-s-last-chance-gulch.html.

3   "Andrew Haraseth (1859–1945)," Find a Grave, https://www.findagrave.com/memorial/8368302
/andrew-haraseth.

4   Amy Sisk, "Numerous Coal, Oil Bills Clear Legislature," *Bismarck Tribune*, May 1, 2021. https://
bismarcktribune.com/news/state-and-regional/govt-and-politics/numerous-coal-oil-bills-clear
-legislature/article_7a4b9d08-c8b5-54c2-a0df-47031529d901.html.

5   James MacPherson, "As North Dakota Oil Soars, so Does Waste of Natural Gas," *Associated
Press*, May 27, 2019, apnews.com/article/9855f0f8c6f146dbb1ebcc92cca3617a.

6   Allison Stegner, interview with the author, July 27, 2021.

7   Rebecca Hersher, "Drawing a Line in the Mud: Scientists Debate When 'Age of Humans'
Began," *NPR*, March 17, 2021. www.npr.org/2021/03/17/974774461/drawing-a-line-in-the-mud
-scientists-debate-when-age-of-humans-began.

8   David Biello, *The Unnatural World: The Race to Remake Civilization in Earth's Newest Age* (New
York: Scribner, 2017), 66.

9   Biello, 4–5.

CHAPTER 19: WE COULD BE RICH

1   Christopher M. Matthews and Collin Eaton, "Harold Hamm, Fracking Pioneer, Faces a
Career Reckoning," *Wall Street Journal*, May 21, 2020, www.wsj.com/articles/harold-hamm
-fracking-pioneer-faces-a-career-reckoning-coronavirus-shutdown-11590074165.

2   Katherine Dunn, "'Unreal': Oil Prices Go Negative for the First Time in History," *Fortune*,
April 20, 2020, fortune.com/2020/04/20/oil-prices-negative-crash-price-crude-market/.

3   Patrick Springer, "'Staggering' 450,000-Barrel Drop in North Dakota's Oil Production as 6,800
Wells Idled," *Inforum*, May 6, 2020, www.inforum.com/business/energy-and-mining
/6480709-Staggering-450000-barrel-drop-in-North-Dakotas-daily-oil-production-as-6800
-wells-idled.

4   Erika Bolstad, "Hope Outlasts Prosperity in Town Flattened by Oil Bust," *E&E News*,
July 27, 2020, https://subscriber.politicopro.com/article/eenews/2020/07/27/hope-outlasts
-prosperity-in-town-flattened-by-oil-bust-012728.

5   Christopher Helman, "Oil Tycoon Harold Hamm Loses $1B Overnight Amid Stock Selloff,"
Forbes, February 27, 2020, www.forbes.com/sites/christopherhelman/2020/02/27/oil-tycoon
-harold-hamm-loses-1b-overnight-amid-stock-selloff/?sh=44f8c11c2e5a.

6   Christopher M. Matthews and Collin Eaton, "Harold Hamm, Fracking Pioneer, Faces a Career Reckoning," *Wall Street Journal*, May 21, 2020, www.wsj.com/articles/harold-hamm -fracking-pioneer-faces-a-career-reckoning-coronavirus-shutdown-11590074165.

7   Renée Jean, "Dueling Opinions on Oil Waste from Continental and the North Dakota Petroleum Council," *Williston Herald*, June 26, 2021, www.willistonherald.com/news /coronavirus/dueling-opinions-on-oil-waste-from-continental-and-the-north-dakota -petroleum-council-at-wednesdays/article_3086fffa-9af4-11ea-8cdd-fb0adccaf7d7.html.

8   Meghan Gordon, "As North Dakota Considers Oil Waste, Only Continental Urges State Output Limits," *S&P Global Platts*, May 20, 2020, www.spglobal.com/platts/en/market -insights/latest-news/natural-gas/052020-as-north-dakota-considers-oil-waste-only -continental-urges-state-output-limits.

9   Prosperity in Town Flattened by Oil Bust," *E&E News*, July 27, 2020, https://subscriber .politicopro.com/article/eenews/2020/07/27/hope-outlasts-prosperity-in-town-flattened-by -oil-bust-012728.

10   Erika Bolstad, "In Slumping Energy States, Plugging Abandoned Wells Could Provide an Economic Boost," *Stateline*, September 23, 2020, www.pewtrusts.org/en/research-and-analysis /blogs/stateline/2020/09/23/in-slumping-energy-states-plugging-abandoned-wells-could -provide-an-economic-boost.

11   "Tribune editorial: Restoring well sites should be ongoing effort," *Bismarck Tribune*, August 30, 2020, https://bismarcktribune.com/opinion/editorial/tribune-editorial-restoring-well-sites-should -be-ongoing-effort/article_889597fc-adab-5b0b-952c-662407d4003f.html.

12   Nicholas Kusnetz, "North Dakota, Using Taxpayer Funds, Bailed Out Oil and Gas Companies by Plugging Abandoned Wells," *Inside Climate News*, May 21, 2021, insideclimatenews.org /news/23052021/north-dakota-orphaned-abandoned-oil-gas-wells-methane-emissions/.

CHAPTER 20: WINDFALL

1   Bridget Reed Morawski, "200-MW NextEra Wind Farm Approved After Wildlife Impact Mitigation Concessions," S&P Global Market Intelligence, June 11, 2020, www.spglobal .com/marketintelligence/en/news-insights/latest-news-headlines/200-mw-nextera-wind-farm -approved-after-wildlife-impact-mitigation-concessions-59015854.

2   Derrick Braaten, phone interview with the author, August 20, 2021.

3   Terry Tempest Williams, "Keeping My Fossil Fuel in the Ground," *New York Times*, March 29, 2016, www.nytimes.com/2016/03/29/opinion/keeping-my-fossil-fuel-in-the-ground.html.

4   Patricia Nelson Limerick, *The Legacy of Conquest* (New York: W.W. Norton & Co., 1987), 71.

INDEX

ABOUT THE AUTHOR

ERIKA BOLSTAD IS A JOURNALIST AND documentary filmmaker in Portland, Oregon. Her work has appeared in the *Washington Post*, *Scientific American*, and many other publications.